A PRIMER OF LIFE HISTORIES: ECOLOGY, EVOLUTION, AND APPLICATION

A Primer of Life Histories

Ecology, Evolution, and Application

Jeffrey A. Hutchings

*Professor of Biology, Department of Biology,
Dalhousie University, Canada*

OXFORD
UNIVERSITY PRESS

OXFORD
UNIVERSITY PRESS

Great Clarendon Street, Oxford, OX2 6DP,
United Kingdom

Oxford University Press is a department of the University of Oxford.
It furthers the University's objective of excellence in research, scholarship,
and education by publishing worldwide. Oxford is a registered trade mark of
Oxford University Press in the UK and in certain other countries

Published in the United States of America by Oxford University Press
198 Madison Avenue, New York, NY 10016, United States of America

British Library Cataloguing in Publication Data
Data available

Library of Congress Control Number: 2021934828

ISBN 978–0–19–883987–3 (hbk.)
ISBN 978–0–19–883988–0 (pbk.)

DOI: 10.1093/oso/9780198839873.001.0001

Printed in Great Britain by
Bell & Bain Ltd., Glasgow

Links to third party websites are provided by Oxford in good faith and
for information only. Oxford disclaims any responsibility for the materials
contained in any third party website referenced in this work.

To my parents, Wendy and Alexander Hutchings

Preface

Science is a way of seeing the world through description, analysis, and interpretation of empirical patterns and processes. Many of the mechanics of science are technical in nature. One may need to know how to run a model simulation, isolate DNA, set a trap, raise seedlings, use a centrifuge, operate a boat engine. The technical demands of research are often obvious.

Less obvious is the need to identify a contextual, interpretative, and analytical framework that allows you to make sense of research findings and to draw conclusions about their potential utility or significance. Are there overarching principles, theories, or other generalities that would aid you in interpreting your research, communicating it to others, and increasing the probability that your work will in some small or large way advance knowledge and understanding? As a *naïve* master's student in the early 1980s, I struggled to identify such a framework. I had little confidence in my ability to distinguish fundamentally important from fundamentally mundane questions.

My confidence received an unexpected boost by Stephen Stearns' 1976 review on life-history evolution, written when he was a graduate student at University of British Columbia. It offered a fresh, taxonomically broad way of thinking about adaptation and natural selection. Why, indeed, should an organism reproduce once in its life and die immediately thereafter?

By 1992, sufficient life-history data were available that allowed for the testing of ideas, the poking of model assumptions, and the prodding of hypotheses. The stage was thus set for the first two general books on life-history evolution. Stearns wrote one; Derek Roff of McGill University wrote the other. Although bearing the same title (*Evolution of life histories*), there were differences in how topics were approached. Reaction norms figured prominently in Stearns' contribution; quantitative genetics was emphasized by Roff. Joined by Roff's *Life history evolution* (2002), these works contributed immeasurably to the torrid pace of life-history research that continues unabated (Figure 1).

Given this apparent enthusiasm, it seemed an opportune time to engage and hopefully enthuse new generations of students and researchers on the grandeur of life-history evolution, its theoretical underpinnings, and some practical applications. Comprising ten chapters, this primer is intended to be accessible to readers from a broad range of academic backgrounds and experience who have interests in ecology, evolution, conservation, or resource management.

Chapters 1 to 4 focus on core elements of life-history theory: population growth; trait variability; trade-offs; genetic architecture; reaction norms; reproductive effort; and reproductive costs. Chapter 5 offers tractable means of estimating fitness and predicting optimal changes in life history, using life tables. The next three chapters examine life-history evolution in variable environments, including bet-hedging (Chapter 6), theories

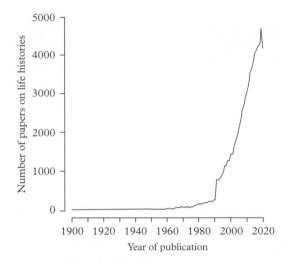

Figure P.1 *Annual number of published papers whose titles, abstracts, or keywords included 'life history' or 'life histories' (hyphenated and non-hyphenated), according to the Web of Science, from 1900 to 2020.*

for the evolution of offspring number and size (Chapter 7), and alternative reproductive tactics and strategies (Chapter 8).

Chapters 9 and 10 bridge the fundamentals of life-history theory to matters of applied interest from conservation and resource-management perspectives. Scaling up from individuals to species, Chapter 9 illustrates how life histories are inextricably linked to the vulnerability of species to extinction, exploitation, and climate change. Chapter 10 completes the primer with a look at how life histories affect sustainable rates of exploitation and how exploitation can, in turn, affect life histories.

The inspiration of this book is large enough. If it fails in its portrayal, the fault lies with an art that is deficient rather than an enthusiasm that is wanting.

Jeffrey A. Hutchings
Halifax, Nova Scotia, Canada
8 January 2021

Acknowledgements

This primer had its genesis on a snowy winter evening in Skåne, Sweden, on 27 January 2014, the result of a lengthy conversation with Per Lundberg, professor of theoretical and evolutionary ecology at Lund University. Four years later, having decided I could create sufficient time to write, I travelled to Iceland to write chapter one in *Auðunnarstofu*, the outstanding fourteenth-century replica building serving as the office of my official host, the Bishop of Hólar, Solveig Lára Guðmundsdóttir. The trip, hosted also by my dear friend and colleague Skúli Skúlason (professor of fish biology at Hólar University), provided me with an ideal setting in which to think, walk, and write. Bjarni Kristófer Kristjánsson (professor and head of the department of aquaculture and fish biology, Hólar University) facilitated my interactions with the Icelandic academic and non-academic communities, never failing to enlighten and inspire me with his cavernous knowledge of Icelandic history. In central Finland, where I wrote considerable portions of the book, I am indebted to Anna Kuparinen (professor of natural resources and environment, University of Jyväskylä) for her patience, support, intellect, and sagacity.

There are many others to whom I wish to offer my sincere thanks, admiration, and appreciation. First and foremost are those who reviewed portions, or the entirety, of one or more draft chapters: Hugues Benoît, Anna Kuparinen, Andrew Simons, Doug Swain, and Laura Weir. Several colleagues provided figures, photographs, data, published works, or unpublished manuscripts during the writing period, including Ken Andersen, Spencer Barrett, Eric Charnov, Larry Greenberg, David Hardie, Richard Law, Susan McRae, Julian Olden, Fanie Pelletier, Jeremy Prince, William Le Quesne, and Patrick White. I am very grateful to Jon Tremaine (Cornwall, UK) whose species-infused hare adorns the book cover. Ian Sherman and Charles Bath (Oxford University Press) faithfully and professionally advised me throughout the writing period.

I wish to acknowledge those who, perhaps unwittingly, provided mentorship and guidance as I strived to educate myself on matters pertaining to life-history evolution through the 1980s and 1990s: Graham Bell, Richard Law, Ransom Myers, Linda Partridge, Derek Roff, and Robert Wootton. Lastly, I owe my greatest thanks to my PhD supervisor Douglas Morris, not only for introducing me to Stearns' (1976) review in 1983, but for impressing upon me the importance of intellectual honesty, academic integrity, respectful interactions with others, and asking questions of fundamental importance.

Contents

1

Fundamentals

1.1   A Brief History of Life Histories

1.1.1   Breadth

Two words are at the core of this book: 'life' and 'history'. Considered singly, writing a primer on either would be presumptuous. But when combined, they offer the foundation of intellectually rewarding, scientifically tractable avenues of inquiry that are not specific to a particular taxon (plant or animal), biological scale (gene, genome, individual, population, species), or mechanism of change (physiology, development, plasticity, evolution).

To be intellectually engaged in the study of life histories from an ecological and evolutionary perspective is to be intellectually engaged in *breadth*. Not all scientists are comfortable with breadth. Some consider it unwise or unhelpful to stray from the comforts of a constrained set of theoretical and empirical constructs that anchor many research programmes. Others find breadth liberating, if not vital to maintaining their intellectual engagement in science over a 30- or 40-year career. If you are used to thinking broadly, tackling a narrow research question can be approached with confidence; if you are used to thinking narrowly, tackling a research question of breadth can be daunting.

To get a feel of the inherent breadth of life-history research, consider a simple figure that illustrates how various factors influence individual life histories, along with some of the consequences these can have at the population level (Figure 1.1). Rather than a mechanistically or formatively accurate flow diagram, think of this figure as a roughly organized cork board into which various constituent elements of life histories have been pinned. Note that none of the elements is specific to a particular taxonomic group; this permits breadth of interest and breadth of inquiry. The box on the left subsamples a range of traits and processes that comprise an individual's phenotype, i.e. its observable form. Following the curved arrow, we are reminded that these traits and processes are influenced, through selection, by an individual's genes, the environment it experiences, and interactions between genotypes and their environment.

The end result is a combination of life-history traits that determines an individual's probability of surviving to, and reproducing at, various ages or stages—its life history. As

A Primer of Life Histories: Ecology, Evolution, and Application. Jeffrey A. Hutchings, Oxford University Press. © Jeffrey A. Hutchings 2021.
DOI: 10.1093/oso/9780198839873.003.0001

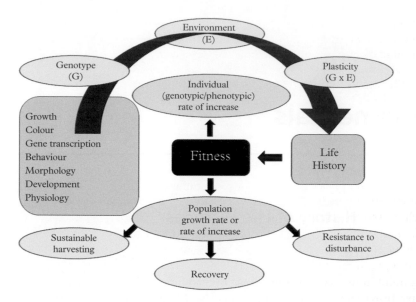

Figure 1.1 *Diagrammatic representation of factors that influence life history which ultimately has direct implications for individual fitness. Correspondingly, individual fitness affects individual and population rates of increase which have consequences for applied issues related to sustainable exploitation and conservation.*

we shall see in this chapter, an individual's life history determines individual fitness (the rate of increase in that individual's genes through time). From an applied and conservation perspective, the average fitness of individuals in a population can affect things such as sustainable rates of exploitation (fishing, forestry, hunting), speed and uncertainty of species/population recovery following depletion, and the ability of a population to persist following natural or human-induced disturbance.

1.1.2 'Life history' in the nineteenth century

The concept of a life history today differs from that of the 1800s when a life history was equivalent to a life cycle. Embryo to egg to larva to adult. Seed to seedling to later vegetative, flowering, and pollination stages. A life history was a description of the development of the presumed generic 'type' of a particular species.

That said, scientific thinking of life histories in the nineteenth century was not entirely devoted to descriptive summaries alone. Charles Darwin (1859) and Ernst Haeckel (1866) were among those who thought evolution to be involved insofar as it could affect things such as the length and number of life stages and their associated morphological and developmental features. After recognition of Georg Mendel's work on plant hybridization (1856–1863) at the turn of the twentieth century, scientists began to explicitly link life histories to genetics. As one example, in 1909 Adam Sedgwick (whose great-uncle of the same name guided Darwin's early studies) asked:

What is the relation of these [developmental] variations in structure, which successively appear in an organism and constitute its life-history, to the mutational variations which appear in different organisms of the same brood or species? (Sedgwick 1909: 181)

As the 1800s eased into the 1900s, life-history studies were very much focused on individual species. There was a paucity of ecology, evolution, and genetics. There was a paucity of breadth. Things were about to change.

1.1.3 The turning point: 1930

Ronald Fisher was pivotal in the development of modern statistical science and mathematical population genetics. Less well-appreciated are his foundational contributions to research on life-history evolution. In the opening chapter of *The genetical theory of natural selection*, he argued that researchers who have accepted the concept of natural selection will 'direct [their] inquiries confidently towards a study of the selective agencies at work throughout the life history' of organisms (Fisher 1930: 21). It marked a pivotal turning point in the development and application of life-history theory.

Fisher bridged the gap between the nineteenth-century concept of a life history and the early twentieth-century concept of natural selection. To do so, he seized a tool long-used in studies of human population growth: 'To obtain a distinct idea of the application of Natural Selection to all stages in the life-history of an organism, use may be made of the ideas developed in the actuarial study of human mortality' (Fisher 1930: 22).

The 'ideas' Fisher was referring to were human actuarial tables. Originally developed in the early 1800s for life-insurance purposes, these tables summarized the probabilities that humans survive from one age to subsequent ages. Fisher went further, arguing that a life table of survival was, in itself, 'inadequate to express fully the relation between an organism and its environment; it concerns itself only with the chances of frequency of death, and not at all with reproduction' (Fisher 1930: 24). By combining probabilities of age-specific survival with age-specific reproduction, he produced what ecologists and population biologists today would term a 'life table'. Fisher showed how one could readily calculate the number of offspring that each newly born individual, or more precisely 'genotype', would be expected to produce over that individual's lifetime.

So, in addition to accounting for the probability of surviving from birth to subsequent ages, Fisher accounted for the rate of reproduction at each age. If each individual in a population produced, on average, sufficient numbers of offspring to exactly replace themselves, the population would remain stable. However, if the per individual or 'per capita' production of offspring surviving to maturity was greater or less than one (i.e. replacement), the population would increase or decline, respectively. Fisher defined the per capita production of offspring as the Malthusian parameter of increase (named after Thomas Malthus whose *An essay on the principle of population* (1798) heavily influenced Darwin and Alfred Wallace).

Fisher argued that the per capita rate of increase should be directly linked to the strength of natural selection. The greater the per capita production of offspring by individuals of a particular genotype, relative to the rate of other members in the same population, the greater the strength of selection.

The necessity of accounting for reproduction when exploring how natural selection acts raised questions concerning the effort an individual should expend on reproduction. What are the consequences to future survival and reproduction associated with the present allocation of greater or lesser amounts of reproductive effort? An oft-quoted sentence from Fisher draws the reader's attention to these potential trade-offs:

> It would be instructive to know not only by what physiological mechanism a just apportionment is made between the nutriment devoted to the gonads and that devoted to the rest of the parental organism, but also what circumstances in the life-history and environment would render profitable the diversion of a greater or lesser share of the available resources towards reproduction. (Fisher 1930: 43–4)

Assuming that the resources available to an individual at any given age are fixed, their diversion to some facet of reproduction must come at the expense of not allocating those same resources to components of survival and future reproductive capacity, such as body maintenance, growth, metabolism, foraging, and vigilance. In other words, present reproductive effort likely comes at a future reproductive cost. Fisher speaks directly to his underlying postulate that the allocation of resources to reproduction, as opposed to other things, is related to 'circumstances in the life-history and environment', i.e. natural selection.

By linking the nineteenth-century, evolutionarily mute, stage-based concept of a life history to his twentieth-century ideas of how natural selection acts (through genetic processes) on age-based probabilities of survival, Fisher brought logical and mathematical clarity to how the strength of selection acting on genotypes varies with age and developmental stage.

A new definition of what constitutes a life history emerged from his work, one that can be expressed in the following way: A life history describes how genotypes vary their age- or stage-specific expenditure of reproductive effort in response to extrinsic and intrinsic factors that affect age-specific survival and fecundity. Or, more succinctly, life histories are the probabilities of survival and the rates of reproduction at each age in a lifespan (Partridge and Harvey 1988).

1.1.4   Fifty years on (1930–1980)

Fisher offered a turning point from nineteenth- to twentieth-century thinking about life-history variability and how natural selection acts on this variation. He laid the foundation for life-history theory. But it was left to others to construct the explanatory and predictive frameworks for understanding why individuals differ so extraordinarily, within and among species, in the means by which they propagate genes to future generations. These efforts were stimulated by fundamental questions of life-history evolution, examples of which are provided as points of departure for cameos that illustrate key developments in the fifty years that elapsed since the publication of Fisher's book.

Question: Why do some organisms breed multiple times, whereas others breed once and then die?

Many organisms reproduce iteratively, or multiple times, throughout their lives. Others breed once and die shortly thereafter. LaMont Cole (1954), who termed these life-history patterns iteroparity and semelparity, respectively, was the first to explore their fitness consequences.

Although Cole (1954) examined several facets of life histories, researchers have been somewhat obsessed with his 'paradox' (sub-section 6.2.1). Surprisingly, he concluded that a semelparous population that produced one offspring more than the mean fecundity or number of offspring (*b*) of an iteroparous population would have the same rate of increase as the iteroparous population. (Put another way, a semelparous population with mean fecundity of (*b* + 1) would have the same rate of increase as an iteroparous population with mean fecundity *b*.) While adding a single offspring to a bird's clutch of five might sound plausible in rendering the fitness of semelparity equal to that of iteroparity, the same cannot be said for adding a single offspring to the million produced by a fish. The paradox is that if an extremely small increase in fecundity should favour semelparity over iteroparity, why is semelparity so uncommon? It turns out that Cole's model incorporated the assumption that iteroparous individuals experience no mortality, not even as very small eggs or seeds. Cole's paradox was solved when empirically realistic rates of juvenile mortality were incorporated and the costs of semelparity were rendered more intuitively reasonable (Charnov and Schaffer 1973; Bell 1976).

Lamentably, Cole's paradox has had a tendency to overshadow his other formative contributions to life-history evolution. A mathematically skilled ecologist (he worked on reptiles), Cole was the first researcher to have overtly concurred with Fisher's perspective on the evolution of life histories, agreeing that 'any life history features affecting reproductive potential are subject to natural selection' (Cole 1954: 104). His classic 1954 paper—*The population consequences of life history phenomena*—was the first since Fisher's book to include the words 'life history' in the title; among the ~310 papers that cited Fisher (1930) between 1930 and 1955, Cole's is the only article to do so (according to Google Scholar).

Question: Why do some organisms start breeding early in life and others comparatively late in life?

The key to answering this question is to understand how age at maturity affects fitness. Cole (1954) explored this issue in detail, demonstrating mathematically that fitness is highly sensitive to the age at which an organism first reproduces, especially at comparatively young ages. All else being equal, the younger you start reproducing, the higher your fitness. But, because of life-history trade-offs, such as the allocation trade-off mentioned in sub-section 1.1.3 (also, see Chapter 4), all else is rarely equal when it comes to comparing the fitness associated with alternative ages at maturity.

Despite the existence of trade-offs, age at maturity retains a dominant influence on fitness. Garth Murphy (1968), who used models of Atlantic herring (*Clupea harengus*) to poke and prod Cole's paradox, concluded that age at maturity depends on the ratio, and the stability, of survival during the juvenile (pre-reproductive) stage relative to survival during the adult (reproductive) stage (sub-sections 5.5.4 and 6.2.2). Murphy's work provided the basis for predicting how changes in survival and fecundity affect the evolution of age at maturity.

Age at maturity is a life-history trait. Cole (1954) may have been the first to explicitly articulate this connection, referring to it as a life-history 'feature.' He also identified fecundity and longevity as life-history features/traits and predicted how they might affect rates of population change. At about the same time, some empirically-minded ecologists were honing in on another life-history trait—offspring size.

Question: Why do some organisms produce many, small offspring and others few, large offspring?

For a fixed amount of 'effort' there is a clear trade-off between the number of offspring produced and the size of each of those offspring, be they seeds, eggs, or embryos. If you are interested in explaining the adaptive significance of variability in one of these two traits (offspring size and offspring number), you will find yourself on intellectually thin ice if you do not account for variability in the other (more of that in Chapter 7).

Early attempts to understand inter-specific variability in offspring number within an evolutionary context can be traced to the ornithologist David Lack whose work in the late 1940s centred on altricial birds (young fed by their parents at a nest). By assuming that the food that parents can feed their young is limited, Lack (1947a) reasoned that the number of young produced cannot increase without a reduction in the amount of food provided to each of those young. In other words, the number of eggs laid reflects the optimal number of young that the parents can feed and/or provide parental care for.

At the same time Lack was pondering the evolution of clutch size in birds, Gunnar Svärdson, a Swedish fish biologist, was considering the evolutionary implications of how offspring size affects offspring number. Foreshadowing what was to come in the early 1970s, he suggested there must be an upper limit to offspring number that depended on how offspring size was related to offspring survival and parental reproductive success (Svärdson 1949). Twenty-five years later, Christopher Smith and Stephen Fretwell (1974), using graphical models, provided an analytical solution for determining the optimal balance between offspring size and offspring number (sub-section 7.4.3). Their simple model remains the starting point today for most explorations of the evolution of egg/seed size.

Question: How costly is reproduction?

The logical necessity of reproductive costs is based on the premise that organisms are energetically constrained systems. If energy required for growth, maintenance, and reproduction originates from the same fixed pool of resources, an allocation of energy to reproduction cannot be made without a reduced allocation to other body functions or activities (Chapter 4).

The genetic basis for reproductive costs was initially explored by Peter Medawar (1952) and George Williams (1957). In discussing the evolution of ageing, or senescence, they argued that natural selection may favour a gene that has beneficial effects early in life even if the same gene greatly reduces fitness later in life. Williams (1966) extended this theory to reproductive effort, reasoning that if energy allocation to present reproduction reduces future reproductive success, this cost of reproduction will result in the evolution of life histories that are characterized by intermediate levels of survival and reproduction.

So, the greater the effort, the greater the subsequent cost. While such a causal link seemed self-evident, it was empirically unclear what this relationship might actually look like. The shape of the function relating reproductive effort to reproductive cost can strongly influence life-history evolution. This is implicit in all theories that are based on the existence of reproductive costs. For example, different shapes of cost functions can lead to the evolution of either iteroparity or semelparity or some combination of the two (Gadgil and Bossert 1970; Schaffer and Rosenzweig 1977; Bell 1980). Although cost functions are central to life-history theory, few have been empirically described. Measuring the magnitude of reproductive costs remains a significant empirical challenge.

1.1.5 Life-history 'classics'

The preceding sub-sections offered a brief history of the core elements of life-history theory to 1980, by which time the foundations of modern approaches had essentially been set. Brevity inevitably involves exclusion and some life-history 'classics' were not mentioned. The interested student and researcher might wish to delve deeper into these foundational papers and a small sample of other fundamentally instructive texts published after 1980 (Table 1.1). (Be it art, music, literature, or science, what constitutes a classic is in the eye of the beholder.)

Table 1.1 *Life-history topics and suggested foundational literature sources.*

Topic	Foundational papers or books
Semelparity vs iteroparity	Cole (1954); Murphy (1968); Gadgil and Bossert (1970); Charnov and Schaffer (1973); Schaffer (1974a, b); Bell (1976); Schaffer and Rosenzweig (1977); Young (1981)
Age at maturity	Alm (1959); Hamilton (1966); Murphy (1968); Stearns (1976); Charlesworth (1980); Roff (1984); Stearns and Koella (1986)
Reproductive effort	Tinkle (1969); Hirshfield and Tinkle (1975); Pianka and Parker (1975); Charlesworth and León (1976); Schaffer and Rosenzweig (1977); Goodman (1984); Charnov et al. (2007)
Offspring size and number	Lack (1947a, b); Svärdson (1949); Cody (1966); Harper (1967); Janzen (1969); Harper et al. (1970); Smith and Fretwell (1974); Brockelman (1975); Capinera (1979)
Costs of reproduction	Medawar (1952); Williams (1957, 1966); Law (1979); Michod (1979); Bell (1980); Rose and Charlesworth (1981); Reznick (1985)
General	Fisher (1930); Lewontin (1965); Williams (1966); Stearns (1976, 1992); Charlesworth (1980); Partridge and Sibly (1991); Roff (1992, 2002)

1.2   A Primer of Population Growth

1.2.1   Intrinsic rate of natural increase

Fisher (1930) placed mathematical population biology firmly at the forefront of scientific research on life-history evolution. Yet, as vital as mathematics has been to the development of life-history theory, the tendency for many biologists—young and old—to eschew quantitative analytics might well have served (and continue to serve) as a retarding force in life-history research, particularly in the empirical testing of life-history theory and its practical applications, of which there are multiple examples (Chapters 9, 10).

As early as the mid-1950s, Cole (1954: 135) was lamenting that studies of life histories have been 'neglected from the evolutionary point of view, apparently because the adaptive values of life-history differences are almost entirely quantitative'. Derek Roff (1992: 3) concluded much the same thing almost 40 years later: 'An early antipathy to the use of mathematical analysis may account in part for the delay in the merging of the ecological and evolutionary perspectives in what is now commonly known as "life history analysis"'.

Fisher clearly did not share this antipathy, embracing Alfred Lotka's (1907; Sharpe and Lotka 1911) model for continuous population growth and applying it within a life-history context. In a closed population (no immigration or emigration) growing at discrete time intervals (all births and deaths occurring at the same time every year), the number of individuals at time step $t + 1$ (N_{t+1}) can be expressed as the number of individuals at time t (N_t) plus the number of individuals born at time t (*Births*$_t$) minus the number of individuals that died at time t (*Deaths*$_t$):

$$N_{t+1} = N_t + Births_t - Deaths_t \qquad \text{Equation 1.1}$$

When modelling the change in population size from time step t to time step $t+1$, it is generally assumed that the likelihood of an individual giving birth or dying in that time interval will be fairly constant. These are termed the per capita rates of birth (*b*) and death (*d*), such that *Births* = *bN* (a rearrangement gives *b* = *Births/N*) and that *Deaths* = *dN* (and *d* = *Deaths/N*). Substituting these per capita rates into a slightly rearranged Equation 1.1 yields the following expression:

$$N_{t+1} = N_t + bN_t - dN_t \qquad \text{Equation 1.2}$$

which can be rewritten as

$$N_{t+1} = (1 + b - d)N_t \qquad \text{Equation 1.3}$$

The parenthetical term in Equation 1.3 is a constant multiplier of population change. It is usually termed the discrete or finite rate of population growth, simplified as λ, such that:

$$N_{t+1} = \lambda N_t \qquad \text{Equation 1.4}$$

Thus, λ is the proportional rate of change in population size from one discrete time step to the next discrete time step, such that:

$$\lambda = N_{t+1}/N_t \qquad \text{Equation 1.5}$$

Although the use of discrete time intervals can be empirically defended, insofar as germination/breeding often occurs at a specific time each year for many species, deaths need not be similarly timed, often occurring continuously.

Mathematically, it can be convenient to simplify population growth as a process that occurs continuously, meaning that changes in population size can be modelled as occurring over extremely small intervals of time known as 'instantaneous change'. A change in population size (N) over such an infinitesimally short period of time (∂t) is represented by $\partial N/\partial t$.

In continuous time, as described above, the number of births is a function of population size, the instantaneous per capita birth rate ($b = Births/N$), and the instantaneous per capita death rate ($d = Deaths/N$). Thus, in continuous time, population growth rate is described by:

$$\partial N/\partial t = bN - dN \qquad \text{Equation 1.6}$$

or

$$\partial N/\partial t = (b-d)N \qquad \text{Equation 1.7}$$

Note that this instantaneous rate of population change is controlled by the difference between the per capita rates of birth and death. Lotka (1907: 22) defined this difference as:

$$r = b - d, \qquad \text{Equation 1.8}$$

terming r the 'rate of natural increase per head' or the per capita rate of natural increase. Substituting Equation 1.8 into Equation 1.7 yields the standard model for population growth rate in continuous time:

$$\partial N/\partial t = rN \qquad \text{Equation 1.9}$$

which means that

$$r = 1/N \ \partial N/\partial t \quad \text{or} \quad \partial N/N\partial t \qquad \text{Equation 1.10}$$

Fisher (1930) defined r similarly (as have most life-history researchers since) but he called it the Malthusian parameter of population increase. According to Cole (1954), Lotka was inconsistent in what he called r, variously describing it as the 'true', 'incipient', 'inherent', and 'intrinsic' rate of increase. Cole, among others (e.g. Birch 1948), settled on 'intrinsic rate of natural increase'. This is generally what r has been called since.

The ecological literature on λ and r can sometimes be confusing because the parameters often seem to be used interchangeably. But they do differ in important ways. Lambda (λ) is measure of population growth rate, i.e. the change in population size from one time step to another, whereas r is a measure of *per capita* population growth rate. Lambda can be thought of as the average contribution of each individual alive at time t to the size of the population at time $t + 1$, whereas r is the average contribution of each individual to the rate of change in population size. The two parameters are related to one another: $\lambda = e^r$ and $r = \ln(\lambda)$.

1.2.2   Density-independent population growth

Whatever one calls r, it is quite important to remember that it is the per capita population growth rate. Consulting Equation 1.10, the units of r are 'individuals per individual per unit of time'. By contrast, population growth rate (Equation 1.9), represented in continuous time by $\partial N/\partial t$, has units of 'individuals per unit of time'. Failure to distinguish per capita population growth from population growth is inexcusably common in the scientific literature.

The discrete (Equation 1.5) and continuous (Equation 1.9) models of population growth are both geometric or exponential functions. We describe this exponential growth by an equation that plots changes in population size over time, such that:

$$N_t = N_0 e^{rt}$$

Equation 1.11

where N_0 is the starting population size. Under these circumstances, we find N increasing steadily without bounds (Figure 1.2). This pattern of change in N with t is also called density-independent population growth.

Under density-independent population growth (Figure 1.2), the per capita population growth rate, r, remains constant; it does not change with population size (N). Rather, it remains at a maximal level, r_{max}, the *maximum* per capita population growth rate (dashed line in Figure 1.3). To improve clarity, Equation 1.9 can be better expressed as:

$$\partial N/\partial t = r_{max} N$$

Equation 1.12

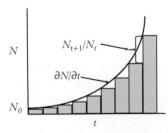

Figure 1.2   *Exponential population growth over discrete time intervals (boxes, illustrating population change once per time step, t) and over continuous time (curved line) from an initial population size of N_0.*

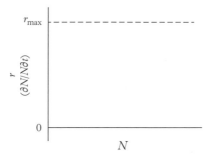

Figure 1.3   *For the density-independent model of population growth, r does not change with population size, N. It remains constant at the maximum value, r_{max}.*

Thus, under exponential growth, the population size at any time t (N_t) depends on two parameters (numbers that remain constant) and one variable (numbers that take on different values), such that N_t depends on the starting population size (N_0, a parameter), the maximum intrinsic rate of increase (r_{max}, also a parameter), and time (t, a variable):

$$N_t = N_0 \exp(r_{max} t) \qquad\qquad \text{Equation 1.13}$$

Equation 1.13 (analogous to Equation 1.11) produces a curve similar in shape to that in Figure 1.2. This basic equation is used to model exponential changes in population size with changes in time. (The phrase $\exp(r_{max} t)$ has come to mean to exponentiate the bracketed term, such that $\exp(x) = e^x$.)

1.2.3   Density-dependent population growth

Under the density-independent model (exponential growth), birth and death rates are assumed to be constant. Indeed, there is good empirical evidence that populations can experience exponential growth when their size is small relative to the numbers of individuals that their environment can sustain, i.e. their carrying capacity (K).

As closed populations increase in abundance or density, their per capita growth rate, r, inevitably changes as competition for increasingly limiting resources, such as space and food, becomes increasingly intense. Now it becomes useful to distinguish 'realized' per capita growth ($r_{realized}$) from maximum per capita growth (r_{max}). Increased competition can have the effect of reducing the realized per capita birth rate and/or increasing the realized rate of per capita death. This means that as N increases, $r_{realized}$ (which equals $b - d$), must decline (Figure 1.4).

The simplest means of incorporating an effect of increasing density (higher N in a closed population) on population growth rate is to reduce $\partial N / \partial t$ by an amount proportional to the remaining 'portion' of the carrying capacity (K), such that:

$$\partial N / \partial t = r_{max} N (1 - N/K) \qquad\qquad \text{Equation 1.14}$$

This equation describes a continuous, density-dependent (or logistic) model in which population growth rate ($\partial N/\partial t$) initially increases with increasing N, reaching a maximum value at $0.5K$, and declines thereafter, eventually reaching 0 when $N = K$ (Figure 1.5).

The pattern of population growth in Figure 1.5 produces a sigmoidal or logistic curve when describing changes in population size over time (Figure 1.6). At small population sizes, density-dependent populations often grow curvilinearly (sometimes exponentially). But as population size continues to increase, the rate of growth reaches a peak at $0.5K$ and then begins to decline, resulting in an ever-decreasing rate as N approaches carrying capacity (Figure 1.6). Note that the slope of the sigmoidal curve in Figure 1.6 represents the population growth rate, i.e. $\partial N/\partial t$.

According to this classic model of density-dependent growth (Equation 1.14), per capita population growth attains its maximum (r_{max}) at the lowest viable population size, declining linearly with increasing N until $N = K$ (solid line in Figure 1.4). The pattern of declining $r_{realized}$ with declining abundance is called negative density dependence.

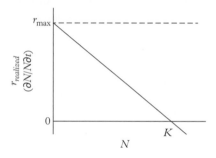

Figure 1.4 *For the density-independent model of population growth, r does not change with population size, N; it remains constant at the maximum value, r_{max} (dashed line). However, for the density-dependent case, $r_{realized}$ declines with N (solid line), falling below zero (and becoming negative) when N exceeds carrying capacity, K.*

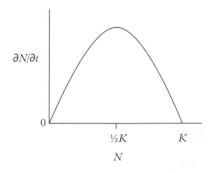

Figure 1.5 *Dome-shaped relationship between population growth rate ($\partial N/\partial t$) and population size (N) for the density-dependent growth model of $\partial N/\partial t = r_{max} N (1 - N/K)$. Maximum population growth rate occurs when population size is half of carrying capacity, K.*

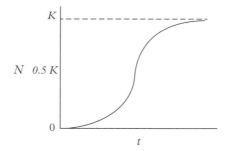

Figure 1.6 *Populations subjected to density-dependent growth increase over time (t) from small population size (N) to carrying capacity (K) in accordance with a sigmoidal or logistic curve.*

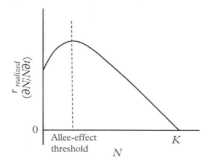

Figure 1.7 *In some species, if population size N falls below a certain level (the Allee-effect threshold), $r_{realized}$ declines as populations become increasingly smaller. This curvilinear pattern of $r_{realized}$ and N is called an Allee effect. Below the Allee-effect threshold, the pattern is indicative of positive density dependence because $r_{realized}$ is positively associated with N at these relatively small population sizes (i.e. small relative to carrying capacity, K).*

Some species, however, experience positive density dependence at small population sizes. Under these circumstances, if populations fall below an abundance or density threshold, $r_{realized}$ begins to decline as populations become increasingly smaller (Figure 1.7). This results in a range of population sizes at which $r_{realized}$ increases, rather than declines, with increasing abundance (producing 'positive' density dependence). The pattern of positive density dependence between $r_{realized}$ and population size is termed an Allee effect (sub-section 9.6.2).

1.3   Life-History Traits

In laying a basic foundation for understanding the origins of life-history research, this chapter has identified four core elements. Two relate to the timing of changes in survival and fecundity throughout an individual's life. These are age-specific survival, l_x, and age-specific fecundity, b_x. The third relates to the concept of reproductive effort, a measure of the investment or energy that individuals allocate to all aspects of reproduction. This comes at some reproductive cost—the fourth core element of a life history—likely to be manifest, at a minimum, as a reduction in the probability of surviving to future ages.

As noted previously, a life history describes how genotypes vary their age- or stage-specific expenditure of reproductive effort in response to extrinsic and intrinsic factors that affect age-specific survival and fecundity. Thus, life-history theory provides an explanatory and predictive framework for understanding why organisms differ in the means by which they propagate their genes to future generations.

This raises the question of what constitutes a life-history trait. Cole (1954) defined a life-history trait as a life-history feature that affects reproductive potential and is subject to natural selection. He identified age at maturity, fecundity (i.e. number of offspring), and lifespan as being life-history traits. To these three, most contemporary life-history researchers would add size at maturity, size of offspring or propagule, and individual growth rate (e.g. Roff 1992, 2002; Stearns 1992).

Other general attributes have been identified as being life-history traits. One of these is reproductive effort. As will be discussed in Chapter 4, one potential measure of effort is the proportional allocation of body mass to reproductive tissues, including offspring mass. This has the advantage of being a quantitatively tractable measure that can be compared among organisms across diverse taxa. However, by restricting the definition to the proportional allocation of body mass to reproductive tissue, one risks overlooking other potential determinants of reproductive effort (such as mate competition, migration, and parental care) and ignoring or underestimating potential costs (for example, to future survival).

A second broadly applicable attribute, sometimes identified as a life-history trait, is offspring sex ratio. Deviations from a 1:1 ratio are exhibited by dioecious plant species, male-biased flowering sex ratios often being twice as common as female-biased ratios (Field et al. 2013). In some animals, individuals are able to vary the proportion of their offspring that are male and female (Charnov 1982; West et al. 2002); in others, they are not (Zietsch et al. 2020).

Other proposed life-history trait candidates are either readily captured by b_x and l_x (e.g. semelparity/iteroparity) or are developmentally taxon-specific (e.g. weaning and gestation periods in mammals; seed dormancy and dispersal in plants; host plant selection by butterflies).

1.4   A Conceptual Life-History Framework

As a means of introducing some of the topics that will be discussed in this book, consider a basic conceptual framework of how selection acting on one life-history trait—age at maturity—can affect the expression of other life-history traits and attributes.

Begin with the premise that extrinsic sources of mortality represent a primary driver of evolution by natural selection. Extrinsic mortality is driven by factors external to an organism, such as predation and disease, as opposed to intrinsic mortality resulting from the consequences of life-history 'decisions', such as the amount of effort to allocate to reproduction.

Figure 1.8(a) illustrates a decline in age-specific survival (l_x) caused by extrinsic factors. Representing the probability of surviving from birth to each age x, l_x is characteristically described by a declining curvilinear relationship. Figure 1.8(a) also distinguishes two periods of life: one that precedes the onset of maturity (the juvenile period) and one that follows maturity (the adult period). The age at which the juvenile period ends and the adult period begins is the age at maturity, α. In many species, the juvenile and adult periods are not constrained to be of fixed duration (they are developmentally flexible), meaning there is variability in the potential lengths of both (represented in the figure by overlapping horizontal lines). This variability, reflected by the orange-coloured region in Figure 1.8(a), means that there are several potential ages at maturity that natural selection can act on (evolution of age at maturity is discussed in Chapters 5, 6, and 8).

Selection on age at maturity often has consequences for size at maturity ($size_\alpha$), especially in organisms that continue to grow throughout their lives. The asymptotic curve in Figure 1.8(b) represents a typical growth pattern, especially in ectotherms (this 'von Bertalanffy' growth curve is discussed in Chapter 2). Size and age at maturity, in turn,

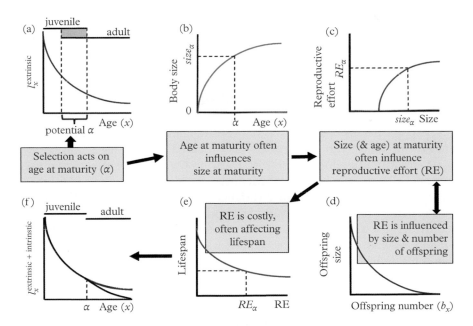

Figure 1.8 *An illustrative example of interactions among core elements of an organism's life history. (a) Mortality from extrinsic sources drives the relationship (red curve) between age-specific survival (l_x) and age (x), providing a mortality framework on which natural selection can act on age at maturity (α). Age at maturity, which distinguishes the juvenile and adult periods of life, has potential to vary within populations, as indicated by the orange colour. (b) The α favoured by selection influences size at maturity ($size_\alpha$) which has consequences for (c) reproductive effort at maturity (RE_α). This effort, reflected in part by (d) the size and number of offspring (age-specific fecundity, b_x), is costly for future survival, thus (e) reducing lifespan. This cost—an intrinsic source of mortality—further reduces the realized values of l_x after maturity (f).*

usually influence the allocation of effort expended on reproduction (RE_α) (Chapter 4). There are multiple ways in which effort can change with age, including an asymptotic relationship (Figure 1.8(c)).

Although reproductive effort typically comprises many elements, two that are central to all organisms are the size and the number of offspring or seeds (Figure 1.8(d); the evolution of offspring size and number is discussed in Chapter 7). When reproductive effort is constrained, such that only a fixed amount of an individual's energy or mass is available for offspring production, there will be a trade-off between offspring size and number, such that one cannot increase without a decline in the other (for seeds and eggs, whose shapes are often roughly spherical, the trade-off takes the form of a negative curvilinear function, the red curve in Figure 1.8(d)).

As elaborated upon in Chapter 4, reproductive effort comes at a cost to future survival, future reproductive success, or both. Figure 1.8(e) illustrates the shape of one type of cost function that can exist between current reproductive effort and lifespan (the greater the effort, the shorter the lifespan).

Lastly, we return to the relationship between age-specific survival and age (Figure 1.8(f)). The black curve incorporates mortality from both extrinsic and intrinsic sources, the latter being driven in this example by a survival cost of reproduction (Figure 1.8(e)). It is the pattern of l_x represented by the black function in Figure 1.8(f) that would correspond to life-table values of age-specific survival (Chapter 5).

1.5 Summing Up and a Look Ahead

In many respects, the history of life-history research mirrors that of evolution by natural selection. In the nineteenth century, the field focused on life stages and life cycles—natural history with a developmental twist. Darwin's *On the origin of species*, coupled with an appreciation of the significance of Mendel's genetic work, created novel theoretical frameworks for interpreting biological variability, one of which lead to Fisher's implicit redefinition of a life history in the 1930s. His work provided the theoretical and mathematical constructs for a key assumption of life-history evolution: natural selection favours those genotypes whose age-specific schedules of survival (l_x) and fecundity (b_x) generate the highest maximum per capita rate of increase (r_{max}) relative to other genotypes in the same population. Cole, Fisher's leading disciple, was at the forefront of a slew of quantitative life-history models that populated the literature through the remainder of the twentieth century.

The perspective of life-history evolution introduced by Fisher and Cole was, at its core, quantitative. The essentials originated in models of population growth by mathematicians such as Lotka. Underlying these mathematical efforts, of course, was the life-history variability that researchers wished to describe, to compare, to understand, and ultimately to predict. This is the subject of Chapter 2.

2

Life-History Variation

2.1   Why Are There So Many Kinds of Life Histories?

Few things in the biological world differ as much as life histories. A 3-mm black fly (Simuliidae) produces hundreds of miniscule eggs within two weeks of hatching, completing its life in less than one month (Adler and McCreadie 2019). Female polar bears (*Ursus maritimus*), weighing several hundred kilograms, mature at four to five years of age and produce one to two cubs, each weighing slightly less than 1 kg, every three years or so (COSEWIC 2018). Few live longer than 20–25 years, the age at which rougheye rockfish (*Sebastes aleutianus*) first begin to spawn at about half a metre in length, producing tens of thousands of small offspring (Haldorson and Love 1991; COSEWIC 2007). This marine fish can live more than two centuries, an impressive lifespan, but one easily superseded by whitebark pine (*Pinus albicaulis*) which matures at 30–50 years, producing tens of thousands of 7–11 mm seeds every few years for as long as a millennium (COSEWIC 2010).

Why are there so many kinds of life histories? This is one of the most fundamental questions in biology and one that is central to this book. But before we can comprehensively address this question, we need to go beyond specific examples. We require context. More specifically, what is the fundamental question pertaining to life-history evolution that we wish to address? Is the question likely to demand the study of a phylogenetically narrow or diverse array of species? We also need to articulate what we mean by life-history differentiation. It might be variability within a single life-history trait. Or we might be interested in knowing how one trait covaries with another trait or even a combination of other traits. Clarity in the question being addressed clarifies the utility of alternative approaches.

This chapter begins with a brief overview of life-history trait variability among species at a coarse resolution of phylogenetic affinity before drilling down into variability between classes within a single subphylum (Vertebrata). The chapter then unfolds with examples of how life histories can be strikingly variable among populations within the same species.

Natural selection is thought to play a dominant role in generating variability in life-history traits within and among populations of the same species (Roff 1992, 2002; Stearns 1992). But among species and higher-level taxonomic ranks—such as families,

A Primer of Life Histories: Ecology, Evolution, and Application. Jeffrey A. Hutchings, Oxford University Press. © Jeffrey A. Hutchings 2021.
DOI: 10.1093/oso/9780198839873.003.0002

orders, and classes—a considerable amount of life-history variation can be attributed to 'constraints'; these can be revealed by patterns of trait covariability. Evidence of one type of constraint emerges when species are unable to express trait values that are common in other species. These constraints can be thought of as developmental, structural, physiological, or genetic boundaries that hinder or limit a species' life-history expression.

A second type of constraint is evident not because of an absence of trait covariation, but rather because of the nature of that covariation, and what it can potentially say about *constancy*. Here, changes in the value of a trait x are constrained to be associated with changes in the value of a trait y in such a way that the division of one by the other produces a constant or invariant value. These are termed life-history invariants.

The chapter concludes with a consideration of how patterns of life-history trait covariation might evolve. The question here is whether traits covary with one another in ways that are reasonably predictable, empirically defensible, and plausibly adaptive. It is these patterns of covariation that have driven efforts to classify trait combinations in accordance with various continuums of divergence, a well-known one being that which distinguishes r- from K-selection.

2.2   Life-History Variability among Species

2.2.1   Differences across phylogenetically diverse organisms

Life-history traits are remarkably divergent among species. Consider age at maturity. Single-celled organisms such as bacteria start reproducing in a matter of minutes, whereas more than a century can elapse before some plants will flower, such as the giant timber bamboo, *Phyllostachys bambusoides* (Janzen 1976). This difference in age at maturity is about 10^6, i.e. six orders of magnitude or a million-fold difference.

You might wonder if this divergence is extreme. But, across broad taxonomic groups, age at maturity is actually one of the least variable life-history traits. Size at maturity differs by 10^7 when comparing a 1-μm bacterium (*Escherichia coli*; Riley 1999) with a 23-m blue whale (*Balaenoptera musculus;* COSEWIC 2002). Within the plant kingdom alone, propagule size varies more than 11 orders of magnitude (10^{11}), the smallest being orchid seeds weighing only 0.0001 mg and the largest being those produced by coco de mer (*Lodoicea maldivica*), each of which can weigh up to 20 kg (Moles et al. 2005).

Nonetheless, an unduly broad comparison of life histories, such as one between a bacterium and a blue whale, might not always be informative. If your interest lies in being able to distinguish the effects of natural selection on life histories from those caused by phylogenetic constraints, you will probably wish to compare species life histories within single phylogenetic entities, such as a class, a family, or even a single species.

Within the eukaryotes (excluding fungi and protists), the phyla with the greatest number of described species are (in ranked order) the Arthropods, Angiosperms, Molluscs, and Chordates. Among these, Angiosperms and Chordates exhibit the greatest range of life-history trait variability. This chapter will focus on the phylum Chordata; more specifically, on the subphylum Vertebrata. What follows in the remainder of

Table 2.1 *References for life-history data presented in Chapter 2, unless otherwise specified. Reference codes: 1 = Myhrvold et al. (2015); 2 = Oliveira et al. (2017); 3 = de Magalhães and Costa (2009); 4 = www.fishbase.org; 5 = Hutchings et al. (2012).*

Class	Body mass	Age at maturity	Clutch size; fecundity	Egg mass	Lifespan
Aves (birds)	1	1	1	1	3
Mammalia	1	1	1		3
Reptilia	1	1	1	1	3
Amphibia	2	2	2		3
Actinopterygii (teleost or bony fishes)	3	3	4		3
Chondrichthyes (sharks, skates, rays)	3	3	5		3

section 2.2 is a broad comparative perspective of life-history trait variability, using frequency distributions (data sources are given in Table 2.1).

2.2.2 Body size

The chapter opened with the question of why there are so many kinds of life histories. The simplest answer is because there are so many different body sizes. For example, assume that organisms cannot produce offspring larger than themselves. If true (a reasonable assumption), the size of offspring that an organism can produce will be limited by the adult size of that organism. Thus, the greater the range in body sizes, the greater the potential variability in offspring size. Put another way, variability in body size permits variability in offspring size.

Among vertebrates, the smallest (mainly fishes) mature at lengths less than 10 mm (e.g. midget dwarfgoby, *Trimmatom nanus*; Winterbottom 1990). The longest is the blue whale (23 m)—a 10^4 difference. But a comparison between the smallest fishes and the largest mammals is not necessary to yield a 10 000 times difference in body length among vertebrates. Maximum lengths of fishes also differ by four orders of magnitude, ranging from 0.8 cm for stout infantfish (*Schindleria brevipinguis*) to 20 m for the whale shark (*Rhincodon typus*) (Olden et al. 2007).

The spread of these data in fishes can be visually well-represented by frequency distribution plots, such as those in Figure 2.1. These plots also illustrate the challenge that exists in visualizing certain types of data and why a transformation of data can be helpful. Olden et al. (2007) compiled an enormous dataset of maximum observed lengths for 22 800 fish species. The data include species from each class that contains extant species. Figure 2.1(a) is a frequency distribution of all data. It is highly skewed or asymmetric (when compared to, say, a bisymmetrical normal distribution). The skew is so great that very few length classes can be visualized; most are grouped in the 1–100 cm or 101–200 cm class. This is caused by the fact that although most fishes never attain lengths longer than 2 m, a very few do. Of the 22 800 species in Olden et al.'s (2007) data set, 176 can achieve lengths longer than 200 cm (92 actinopterygians and 84 elasmobranchs).

We might try to improve our visualization of the data if we exclude these relatively few (0.77 per cent) large species, as in Figure 2.1(b). This is certainly a visual improvement over Figure 2.1(a), but the data (*n* = 22 624 species) still bear a strong skew. Such highly skewed data cannot be used to calculate basic descriptive statistics of the data, such as the mean, standard deviation, variance, or standard error. To calculate these values, the

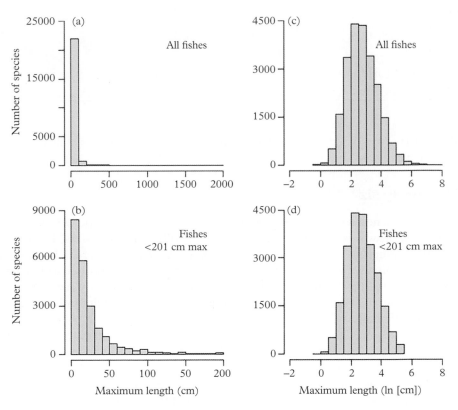

Figure 2.1 *Frequency distribution of maximum recorded body length in fishes. (a) All data (n = 22 800 species); (b) species for which maximum length is less than 201 cm (n = 22 624 species); (c) log-transformed (using ln, or natural logarithm) data for all species; and (d) log-transformed data for species for which maximum length is less than 201 cm. For the data represented in Figures 2.1(a) and (c), the sample sizes of species numbers by taxonomic group are as follows: 22 031 ray-finned fishes (class Actinopterygii); 632 sharks, skates, and rays (class Chondrichthyes, subclass Elasmobranchii); 64 hagfish (class Myxini); 39 lamprey (class Cephalaspidomorphi); 23 chimaeras (class Chondrichthyes, subclass Holocephali); and 11 lobe-finned fishes (class Sarcopterygii).*

data need to approximate a normal distribution. To achieve normality from a highly skewed distribution, the data can be transformed. Here, this is achieved by taking the natural logarithm (ln) of each original data point and then plotting the distribution of these log-transformed data. When using the natural log, rather than plotting x, we are plotting $\ln(x)$.

The frequency distributions of the log-transformed data are shown for all fish species and for those that do not attain more than 200 cm in length in Figures 2.1(c) and 2.1(d), respectively. For all species (Figure 2.1(c)), the mean of the log-transformed data is 2.75 and the standard deviation (σ) and standard errors ($\sigma/\sqrt{n}$) are 1.00 and 0.007, respectively. To convert these log-transformed values to their original values, exponentiate each value (i.e. e^x), such that the mean maximum length, standard deviation, and standard error for all 22 800 species of fishes in the dataset, measured in cm, are 15.64, 2.72, and 1.01, respectively.

As with the fish maximum length data, some of the trait frequency distributions that follow in this chapter would also be highly skewed if the original data were not log-transformed. One such trait is maximum body mass, expressed across six classes (Figure 2.2): Aves (birds), Mammalia, Reptilia, Amphibia, Actinopterygii (ray-finned fishes, including teleost or bony fishes), and Chondrichthyes (sharks, skates, and rays). Although there are multiple ways in which the data could be compared, we will focus on the median (a measure of central tendency, the median is the middle value of a descending or ascending list of numbers) and the coefficient of variation (the CV, a standardized measure of the dispersion or variation of a frequency distribution, is the standard deviation divided by the mean).

Based on estimates of the median from the data used to construct Figure 2.2, mammals are the largest (82.27 g) among the primarily terrestrial vertebrates followed by birds (36.97 g), reptiles (30.27 g), and amphibians (9.30 g). Relative to the mean, the variation in maximum weight is highest among amphibians (CV = 2.18), followed by reptiles (1.95), mammals (1.75), and birds (1.49). Comparing the two groups of fishes, chondrichthyans attain larger maximum weights (median = 45 251.9 g) than teleost fishes (2500.0 g) but slightly smaller CV (Chondrichthyes: CV=1.22; teleosts: CV=1.43).

2.2.3 Age at maturity

Median age at maturity differs seven-fold among classes of vertebrates (Figure 2.3). Mammals and birds mature at the youngest age (1 yr). Teleost fishes (3 yr), amphibians (3 yr), and reptiles (4 yr) mature at three to four times the median ages of birds and mammals. Chondrichthyans mature at the oldest median age (7 yr). In terms of trait variability, CVs are highest among chondrichthyans (1.49) and mammals (1.24), intermediate among teleosts (0.92), birds (0.88), and reptiles (0.80), and lowest among amphibians (0.57). (Note that although some of these distributions are skewed, the degree of skewness is not as severe as in Figure 2.1(a); a log-transformation is not as necessary to visualize the data, although it was used to calculate the CV.)

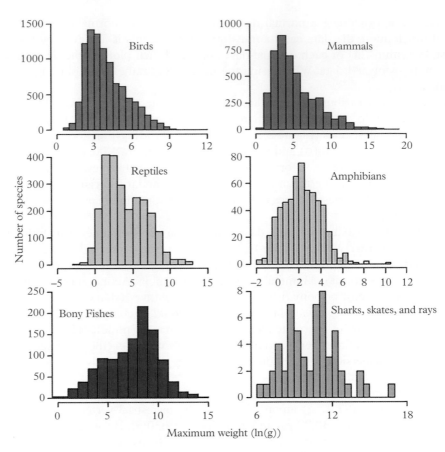

Figure 2.2 *Maximum body size for six classes of vertebrates. Birds (n = 9532 species), mammals (n = 4651), reptiles (n = 2494), amphibians (n = 591), teleosts (bony fishes; n = 1092), and chondrichthyans (sharks, skates, and rays; n = 57). See Table 2.1 for data sources, except data for bony fishes are from www.fishbase.org.*

2.2.4 Offspring number

The seven-fold difference in median age at maturity pales by comparison to the differences in offspring number among vertebrate classes (offspring number or fecundity is often referred to as clutch size or litter size in bird and mammal studies, respectively). While the production of 10 or fewer offspring per breeding event is common in birds, mammals, reptiles, and chondrichthyans, it is rare among amphibians and bony fishes (Figure 2.4). As evident from a visual inspection of Figure 2.4, the median fecundity is highest in bony fishes (9616) and amphibians (400). Reptiles (7.6) and chondrichthyan fishes (7.3) share similar median fecundities, as do birds (3.0) and mammals (2.2). In terms of CV (calculated from log-transformed data), these are remarkably similar among

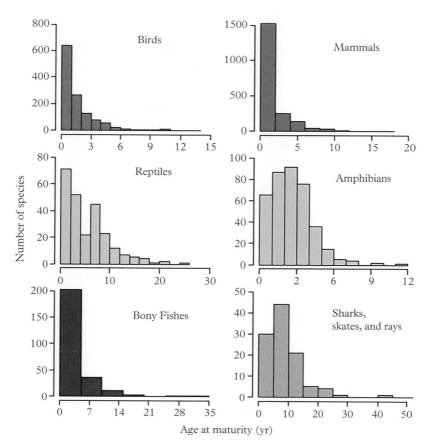

Figure 2.3 *Age at maturity for six classes of vertebrates. Birds (n = 1224 species), mammals (n = 2000), reptiles (n = 245), amphibians (n = 377), teleosts (bony fishes; n = 254), and chondrichthyans (sharks, skates, and rays; n = 107).*

classes; in ranked order, they are mammals (CV=2.30), chondrichthyans (1.70), birds (1.62), bony fishes (1.46), amphibians (1.38), and reptiles (1.35).

2.2.5 Offspring size

Comparing offspring size among vertebrate classes is problematic because of differences in the development of the fertilized egg. In many chondrichthyan fishes and mammals, the hatching of the egg occurs in the oviduct. The subsequent offspring size is, thus, heavily influenced by the length of the gestation period during which the offspring is developing in the uterus. Birds do not bear live young. Nor do most fishes, amphibians, and reptiles. Unfortunately, databases on egg size tend to use different units of measurement. Egg diameter is favoured over egg mass in fish studies (Wootton 1998); egg mass is favoured over egg diameter in bird studies (probably because bird eggs are

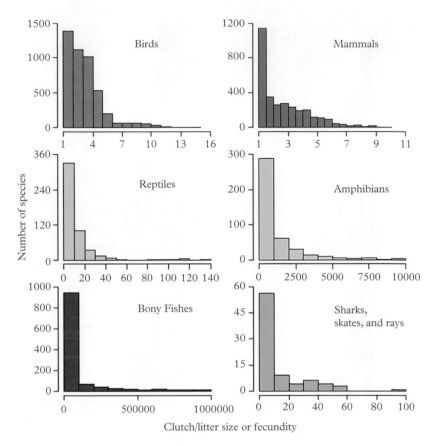

Figure 2.4 *Data on offspring number per breeding event (variously termed clutch size, litter size, or fecundity) for six classes of vertebrates. Birds (n = 4542 species), mammals (n = 3115), reptiles (n = 493), amphibians (n = 425), teleosts (bony fishes; n = 1118), and chondrichthyans (sharks, skates, and rays; n = 107). Data omitted for clarity: birds with clutch size >15; mammals with litter size >10; teleost fishes with fecundity >1 000 000.*

usually not spherical) (Myhrvold et al. 2015); and offspring length appears to be more likely to be recorded than individual egg mass or diameter in studies of amphibians (Oliveira et al. 2017).

Restricting comparisons of egg mass between birds and reptiles (two classes for which extensive data on egg mass are available) (Figure 2.5), reptiles have a much larger median egg mass (11.44 g) than birds (4.53 g) but the standardized variation in egg mass is lower in reptiles (CV=1.45) than it is in birds (CV=2.50). Among the largest eggs produced by teleost fishes are those produced by Chinook salmon (*Oncorhynchus tshawytscha*). After extrusion by the female, each egg is about 500 mg or 0.5 g in mass (Rombough 1985; Kamler 1992). Although the egg sizes produced by birds and reptiles vastly exceed the maximum for fishes (see the two right-hand panels in Figure 2.5), the

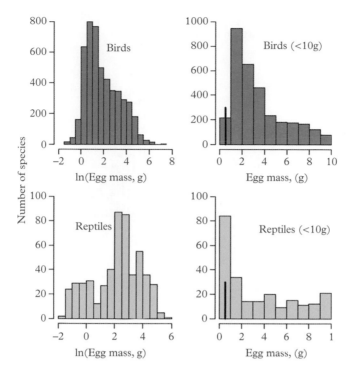

Figure 2.5 *Distribution of ln(egg size, g) for all birds and reptiles in left panels and for birds and reptiles with eggs that weigh less than 10 g (right panels). Black vertical lines at 0.5 g identify the maximum fish egg weight. Species sample sizes: all data (birds: n = 4888; reptiles: n = 527); eggs less than 10 g (birds: n = 3231; reptiles: n = 234).*

eggs of bony fishes are comparable in mass to, albeit generally much smaller than, seeds produced by plants (Figure 2.6).

2.2.6 Lifespan

Maximum recorded age differs considerably among vertebrates (Figure 2.7). Based on their median values, vertebrate classes can roughly be distinguished by three groups. The youngest are teleost fishes (10.0 yr) and amphibians (12.4 yr); the intermediate group includes birds (15.3 yr), mammals (17.1 yr), and reptiles (17.8 yr); and the oldest are the chondrichthyan fishes (19.4 yr). In terms of data variability, the CVs in maximum lifespan of chondrichthyan (1.46) and teleost fishes (1.26) are considerably greater than those of reptiles (0.81), mammals (0.78), amphibians (0.71), and birds (0.62). According to the data compiled by de Magalhães and Costa (2009), the species with the maximum lifespans in each vertebrate class are as follows: chondrichthyan fishes (392 yr in Greenland sharks, *Somniosus microcephalus*); mammals (211 yr for bowhead whales, *Balaena mysticetus*); teleost fishes (205 yr in rougheye rockfish, *Sebastes aleutianus*);

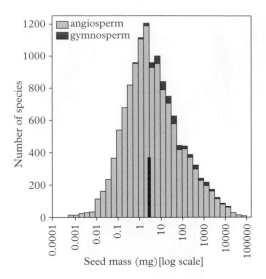

Figure 2.6 *Distribution of seed mass (log$_{10}$(mg)) for 12 987 species of plants. Black vertical line indicates the largest egg mass of a bony fish.*

Source: Moles et al. (2005). Reprinted by permission from the American Association for the Advancement of Science.

reptiles (177 yr in the Galápagos tortoise, *Chelonoidis nigra*); amphibians (102 yr in the olm, *Proteus anguinus*); and birds (83 yr in pink cockatoos, *Cacatua leadbeateri*).

2.2.7   Life-history differences within species can be considerable

Based on the data illustrated thus far, there is clearly a great deal of variability in life-history traits among species within vertebrate classes. But natural selection acts on variation among individuals within populations of a single species. This raises the question of how variable life-history traits can be within single species relative to what is observed among species. A particularly variable vertebrate family in this regard is the fish family Salmonidae (salmons, trouts, chars) for which differences in life history have been attributed to a high degree of plasticity in habitat and migratory behaviour (plasticity is a key topic in Chapter 3), coupled with plasticity in development, physiology, and reproductive strategy (Jonsson and Jonsson 2011; Quinn 2018).

Arguably the most variable of the salmonids, and thus of vertebrates, is the Atlantic salmon (*Salmo salar*). This can be seen by comparing the range in four life-history traits expressed by different populations of salmon with data for the same traits in four non-fish vertebrate classes: Aves, Mammalia, Reptilia, and Amphibia. The range in population

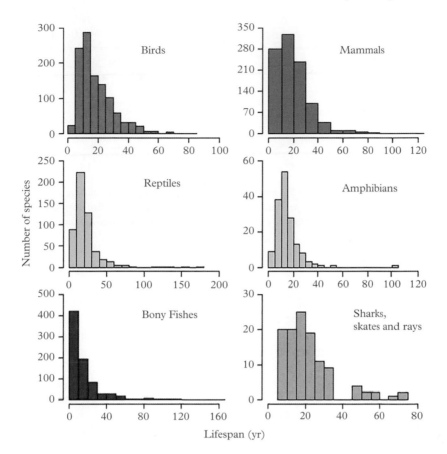

Figure 2.7 *Maximum lifespan for size classes of vertebrates. Birds (n = 1117 species), mammals (n = 1010), reptiles (n = 522), amphibians (n = 158), teleosts (bony fishes; n = 796), and chondrichthyans (sharks, skates, and rays; n = 115). For visual clarity, lifespan data have been excluded for bowhead whale (211 yr), rougheye rockfish (205 yr), and Greenland shark (392 yr).*

differences within Atlantic salmon exceeds the 25 per cent to 75 per cent quantile range for these chordate classes in terms of maximum weight, age at maturity, and fecundity, and comes close to matching that for these chordates in terms of lifespan (Figure 2.8).

At one extreme, female *S. salar* that spend their entire lives in small streams mature at 10 cm (17 g), produce 30–35 eggs per female, and reproduce one to three times throughout their lives; at the other extreme, females that spawn in very long, northern rivers, with access to and from the sea, can mature at lengths greater than 120 cm, attain maximum weights of almost 50 kg (fishbase.org), and produce more than 10 000 eggs per breeding season (Hutchings et al. 2019).

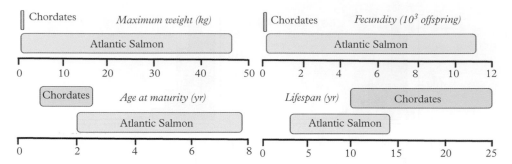

Figure 2.8 *Range in values of four life-history traits among populations of Atlantic salmon, compared to the 25–75 per cent quantile range for data combined for four classes of non-fish chordates (birds, mammals, reptiles, and amphibians). Data sources: Table 2.1, supplemented by information from fishbase.org and Hutchings et al. (2019).*

2.2.8 Growth rate

Growth is determinate or indeterminate. Broadly defined, determinate growers (e.g. mammals, birds) cease growing after maturity whereas indeterminate growers (e.g. plants, fishes, reptiles) continue to grow, albeit at increasingly slower rates, throughout life. Growth has its most obvious effect on body size. Body size is of general importance to multiple life-history traits and parameters, such as lifespan (Figure 2.9) and, within populations of fishes (Wootton 1998) and plants (Silverton et al. 1997), fecundity.

The importance of growth rate to life history differs between growth types, having greater influence on indeterminately growing organisms. There are two key reasons for this. Firstly, growth rate, and selection on growth rate, has potential to change throughout the life of an indeterminate grower. Secondly, indeterminate growers are ectotherms, gaining their heat primarily through the environment. As a consequence, the environment tends to have considerably greater influence on the rate of change in body mass of indeterminate growers than of determinate growers.

Reflecting both individual size at age and the rate at which that size is attained, growth rate can significantly affect life-history traits, thus influencing age-specific rates of survival (l_x) and fecundity (b_x). The traits most generally affected are age and size at maturity. But the direction of these effects is not always readily predictable. Nor can one conclude that faster growth will always result in higher fitness. For example, fast growth early in life (during the juvenile, pre-reproductive period) can increase the fitness benefits of delaying sexual maturity and increasing adult size (Day and Rowe 2002). By increasing adult size, individuals can increase b_x: larger individuals tend to produce more offspring (benefitting females primarily) and can secure more mating opportunities (benefitting males primarily). But in most fishes, faster growth is associated with younger age at maturity, often at a smaller size, than those experiencing slower growth (Alm 1959; Wootton 1998).

Regarding l_x, the larger size at age achieved by faster-growing individuals often leads to lower mortality because of the survival benefits (such as reduced susceptibility to

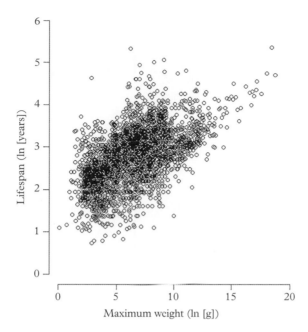

Figure 2.9 *Maximum lifespan increases with maximum body mass among 3702 species of vertebrates: birds (n = 1192), mammals (n = 1329), reptiles (n = 544), amphibians (n = 174), teleost fishes (n = 346), and chondrichthyan fishes (n = 117).*

Data source: de Magalhães and Costa (2009).

predation) associated with larger body size (Shuter and Post 1990; Conover 1992). But these fitness benefits, achieved *after* the larger sizes have been attained, can be countered by costs associated with securing faster growth as juveniles *before* the larger sizes are attained. Rapid attainment of body sizes can come at metabolically related costs to other functions (Arnott et al. 2006). Arendt (1997) provides a taxonomically broad consideration of potential fitness costs and benefits associated with slow and rapid growth.

Thus, there is difficulty in identifying generalities associated with how individual growth affects life history. As already intimated, a primary reason for this difficulty is that one also needs to understand how mortality varies with growth rate. Stearns and Koella (1986) explored how patterns of covariation between age and size at maturity depend on the relationship between individual growth rate and mortality during the juvenile (pre-reproductive) and adult stages of life. Their model outputs illustrate the complexity of predicting how individual growth rate affects life history (Figure 2.10). In other words, the influence of growth rate on life-history traits is not straightforward, often resulting in non-intuitive relationships, as illustrated in Figure 2.10. Thus, while growth rate can be a primary determinant of life-history traits and fitness in some organisms, there are challenges in drawing generalities applicable to all organisms.

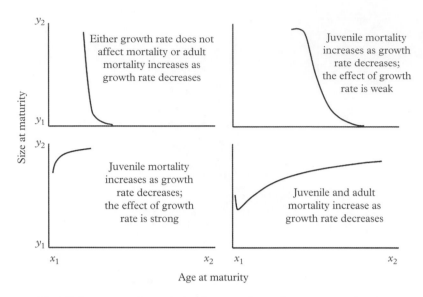

Figure 2.10 *Modelled patterns of covariation between size and age at maturity that depend on the relationship between individual growth rate and individual mortality.*

Source: Re-drawn from Stearns and Koella (1986). Reprinted by permission from John Wiley and Sons.

2.3   Life-History Constraints and Invariants

2.3.1   Constraints

Grandiose comparisons such as those offered at the beginning of this chapter have a sensational element to them. The biggest this; the smallest that. These comparisons are important in what they reveal about finite boundaries in the phenotypic expression of life-history traits. It becomes evident that trait values observed in some taxonomic groups are not observed in others. For example, the smallest recorded egg from among ~10 000 described species of birds (bee hummingbird, *Mellisuga helenae*) has a mass of 0.25 g and diameter of ~7 mm (Myhrvold et al. 2015). This is a very small egg when compared to other birds. But it is larger than the egg size of most species of fishes.

   The minimal overlap in egg size between birds and fishes reflects a constraint. Birds produce larger eggs than fishes because of developmental, biomechanical, physiological, or phylogenetic limitations that prevent them from producing an egg smaller than ~7 mm. Constraints set boundaries or limits on trait expression that can prevent variation in a given trait in one species from overlapping ranges in the same trait in another species. These constraints are likely (but need not be) a consequence of natural selection in the phylogenetically distant past.

   A bivariate plot of fecundity versus body size within the subphylum Vertebrata provides one example of a likely developmental constraint. As Figure 2.11 illustrates, bony fishes realize vastly greater fecundity than birds, mammals, and reptiles combined

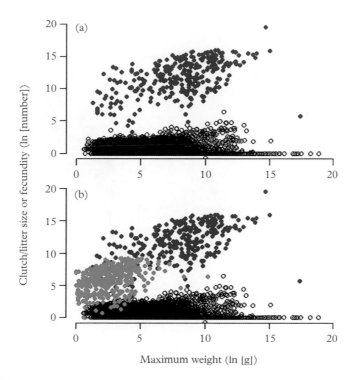

Figure 2.11 *Bivariate plots of offspring number and body mass, distinguishing (a) data for bony fishes (n = 316) (blue) from data for birds (n = 4542 species), mammals (n = 3115), reptiles (n = 493), and chondrichthyans (n = 483) (open circles), and (b) amphibians (n = 391) (green).*

Data sources as in Table 2.1.

(Figure 2.11(a)). It is not because fishes are larger and perhaps have a greater body volume than these other vertebrates. One possibility is that birds, mammals, and reptiles cannot produce nearly as many offspring as fishes because natural selection acts against small offspring size in these classes (Chapter 7). Amphibians bridge some of the empty state space in Figure 2.11(a) (see Figure 2.11(b)), perhaps because they share with fishes an aquatic embryonic environment.

Constraints can also be evident when examining bivariate plots of other traits. The blue-shaded regions of Figure 2.12 identify areas of bivariate space populated by few if any vertebrates, such that large body size is associated neither with young age at maturity nor short lifespan.

2.3.2 Life-history invariants

Data in the previous sub-section were used to identify potential life-history constraints by revealing trait values that do not overlap between species (Figure 2.11(a)) or that are simply absent (Figure 2.12). Another form of constraint is evident when traits covary in

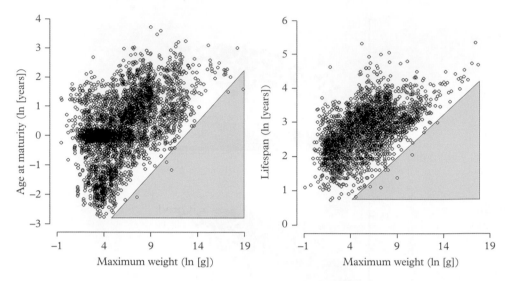

Figure 2.12 *Bivariate plots of age at maturity and lifespan as functions of maximum body mass in vertebrates. The paucity of data in the arbitrarily drawn blue-shaded regions suggests the existence of constraints that prevent large-bodied organisms from maturing at young ages or from experiencing short lifespans.*

such a way that the ratio of trait values remains constant. Put another way, the traits are constrained to vary with one another in such a way that the quantitative relation between the two values remains the same.

In the late 1950s, fisheries scientists Sidney Holt and Ray Beverton began to explore how life-history traits and parameters associated with age, size, growth, and mortality might be related (Holt 1958; Beverton 1963). One of their objectives was to see whether simple relationships between traits that are relatively easy to measure could be used to estimate traits that can be difficult to measure, such as mortality (reflected by M, the instantaneous rate of natural mortality), and age and size at maturity.

Beverton and Holt made use of the output of a very simple (and now universally applied) model to describe individual growth. Developed by von Bertalanffy (1938), the model articulates a curvilinear relationship between size and age in indeterminately growing organisms (Figure 2.13). Two key parameters can be estimated from this model: the 'asymptotic' length (L_∞, also termed '$L_{infinity}$') and the growth coefficient (k) which describes how rapidly or how slowly the asymptotic length is reached. (L_∞ and k are often, albeit imprecisely, described as maximum length and individual growth rate.)

Using data on species from a single order (Clupeiformes: herring, anchovies, and sardines), Beverton and Holt (1959) analysed relationships between M, L_∞, k, age at maturity (α), and length at maturity (L_α). They found that certain combinations yielded constant or invariant ratios. Beverton (1963) expanded the species examined to representatives of three additional orders. Three apparently invariant life-history ratios were identified: $M/k = 1.5$; $M\alpha = 1.65$; and $L_\alpha/L_\infty = 0.66$. These relations of constancy,

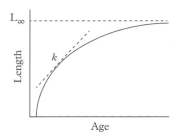

Figure 2.13 *The von Bertalanffy model of individual growth. L_∞ is the asymptotic length and k is the growth coefficient which describes the rate at which L_∞ is reached.*

or life-history invariants, imply that a change in one trait produces a predictable change in another life-history trait. For example, if you have an estimate of k for individuals in a population, multiply it by 1.5 to obtain an estimate of M.

The study of life-history invariants was strengthened and broadened considerably by Eric Charnov (1993) who presented theoretically and empirically compelling arguments that these invariants reflect adaptive processes of a broad and general nature. More specifically, invariance among parameters related to growth, survival, and mortality suggested the existence of universal, optimally adaptive trade-offs resulting from natural selection. The invariant M/k implies that fast-growing individuals experience higher natural mortality rates than slow-growing individuals. The invariant L_α/L_∞ implies that maturity is achieved at a length approximately two-thirds of the maximum.

One question that arises is whether life-history ratios calculated for one group of species retain the same value when the database is broadened to include other groups of species. Expanding the empirical data to 41 fish families (175 populations), Pauly (1980) found general support for Beverton and Holt's results. Upon broader examination, Charnov (1993) updated some invariants for fishes, often specifying ranges rather than single constants: M/k = 1.65 to 2.10; $M\alpha$ = 1.75 to 2.20.

Pauly and Charnov's expansion of the study of invariants drew attention to an emerging caveat. The values of life-history ratios when estimated at one taxonomic level were not always readily transferable to other taxonomic levels. Prince et al. (2015) provide a clear empirical example of this in an analysis of 10 families of marine fishes. They reported that M/k differs among families and is negatively associated with L_α/L_∞ (Figure 2.14). Their work was affirmed by Thorson et al.'s (2017) study of more than 32 000 species of fishes, which also found little evidence of constancy in M/k among families (Figure 2.15).

A second caveat relates to the means by which life-history invariants are calculated. They can be biased because they represent ratios of parameters that are often strongly correlated with, rather than being entirely independent of, one another (Nee et al. 2005; Pardo et al. 2013). Prince et al. (2021) suggest ways of addressing these deficiencies by incorporating standardization and quality-control procedures.

However, the analyses by Prince et al. (2015, 2021) and Thorson et al. (2017) should not be interpreted to mean that life-history invariants do not exist. Rather, they remind

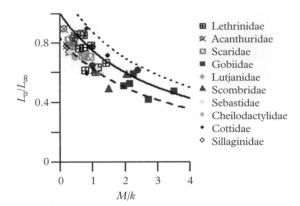

Figure 2.14 *Relationship between two life-history invariants for species from ten families of marine fishes. The invariants are age at maturity (L_α) relative to asymptotic length (L_∞) and natural mortality (M) relative to the von Bertalanffy growth coefficient (k).*

Source: Prince et al. (2015). Reprinted by permission from Oxford University Press.

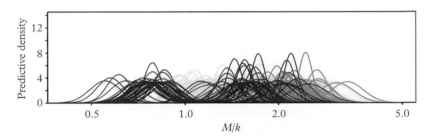

Figure 2.15 *Predictive distributions for the ratio of natural mortality (M) to the von Bertalanffy growth coefficient (k). The estimate of M/k for each species is normally distributed on either side of the model-estimated mean. The species are members of four different families of fishes. Red: Sebastidae (rockfishes, ocean perches). Green: Salmonidae (salmons, trouts, chars). Blue: Scombridae (mackerels, tunas). Yellow: Lutjanidae (snappers).*

Source: Thorson et al. (2017). Reprinted by permission from John Wiley and Sons.

us to be cautious in their application, being mindful of their empirical strengths, weaknesses, and potential biases. From an evolutionary perspective, the evidence for invariance in fishes increases as one analyzes data at increasingly lower taxonomic categories, such as from class to order to family to genus. This might well reflect natural selection for optimal trade-offs. From a practical perspective, when using life-history invariants in vulnerability assessments of data-poor species (see Chapters 9 and 10), the application of life-history ratios, such as *M/k*, can be strengthened by accounting for taxonomic affiliation (Prince et al. 2021).

2.4   Patterns of Trait Covariation

2.4.1   Trait covariation in plants

Sections 2.2 and 2.3 made direct or indirect reference to some common patterns of trait covariation. Larger species, for example, tend to live longer lives and mature at older ages (Figure 2.12). Species that have very high fecundity produce relatively small offspring, a pattern of covariation quite evident in fishes and plants. It has long been assumed that life histories have co-evolved to form highly generalized patterns of co-adapted traits. If true, these patterns should be broadly evident among phylogenetically diverse groups of organisms.

One early and enduring example is a scheme for plants developed by Philip Grime (1977). His approach to classifying primary strategies was based on the idea that each species faces an evolutionary trade-off between competing for resources, enduring resource limitation, and recovering from (naturally occurring) biomass destruction (Grime and Pierce 2012). He proposed that plants could be assigned to one of three groups that differ in life history and habitat: competitors (C), ruderals (R), and stress-tolerators (S). As his classification scheme indicates (Figure 2.16), Grime (1977) was particularly interested in the influence of competition, disturbance, and stress.

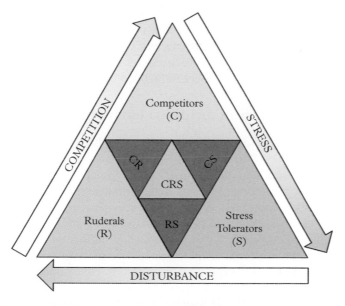

Figure 2.16 *Grime's (1977) classification of plant life histories into Competitors (C), Ruderals (R), and Stress Tolerators (S), based on axes of selection related to competition (C-selection), stress (S-selection), and disturbance (R-selection).*

Competitors are plants that dominate habitats in which environmental stress is relatively benign and disturbance is rare, resulting in competition being a presumed major agent of natural selection. Competitive plants, such as trees and shrubs, are effective at acquiring and consuming resources. They tend to delay maturity until they have attained relatively large sizes, often producing broad leaf canopies and extensive root systems. *Stress-tolerators* are adapted to environments that are difficult in terms of climate, moisture, and nutrients, but that are stable because they are infrequently disturbed. Stress-tolerant plants are typical of extremely hot and cold desert-like environments. Examples include Arctic heather (*Cassiope tetragona*) and cacti (family Cactaceae). They are generally short of stature and intolerant of competition. Their life histories are typified by slow growth, small size, delayed maturity, and long lives. *Ruderal* plants are adapted to living in recently disturbed habitats that have abundant resources. Examples include annual herbs and agricultural weeds. These plants typically grow rapidly, mature early, produce high numbers of seeds with broad dispersal capabilities, and are intolerant of competition and stress.

Grime's (1977) classification scheme has proven to be a remarkably robust framework for classifying plant life histories, based on the degree to which they are subjected to selection resulting from competition (C-selection), stress (S-selection), and disturbance (R-selection) (Pierce et al. 2017) (Figure 2.17). Notwithstanding some inevitable drawbacks (Mahaut et al. 2020), it continues to serve as a basis for understanding how selection can influence the life histories of plants and other organisms. For example, conceptual models and genetic studies suggest that fungal-mediated plant decomposition rates might be governed by a trade-off between stress tolerance and competitive dominance (Lustenhouwer et al. 2020). Grime's (1977) scheme has also served to stimulate new classification schemes, such Archibald et al.'s (2019) framework for understanding how plant life histories are influenced by fire and herbivory.

2.4.2   Trait covariation in animals: early thinking

Evolutionary biologists Theodosius Dobzhansky and Ivan Schmalhausen have been attributed with early thinking on how co-adapted life-history traits might be expressed. Dobzhansky (1950) asked how an organism's environment might affect natural selection, particularly in terms of the probability that a population would fluctuate because of regular episodes of mass mortality, as opposed to experiencing relative stability near an equilibrium, such as when a population is near its carrying capacity (K). He postulated that species that live in physically harsh conditions in temperate (especially Arctic and montane) environments are more likely to experience massive bouts of mortality than those in the tropics because of unpredictable environmental change, such as excessive cold or drought.

Citing Schmalhausen (1949), Dobzhansky predicted that regular, high-mortality events would favour species that have increased fecundity and accelerated development and reproduction. In contrast, in the tropics, 'where physical conditions are easy, interrelationships between competing and symbiotic species become the paramount adaptive

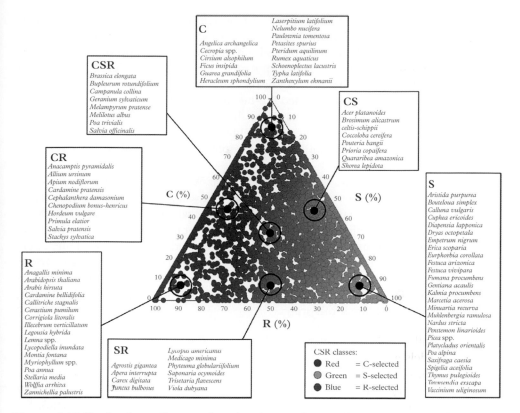

Figure *2.17 Tracheophytes (i.e. vascular plants), according to Grime's (1977) classification (see Figure 2.16), based on an assessment of the degree to which each species is subjected to C-, S-, and R-selection.*

From Pierce et al. (2017). Reprinted by permission from John Wiley and Sons.

problem' (Dobzhansky 1950: 217), organisms should be favoured to adopt life histories that would enhance their competitiveness.

Although Dobzhansky does not mention the word 'density', it is fair to say it is implicit in his arguments. Species subjected to regular, mass-mortality events are less likely to be affected by the competition associated with high population density than species that are not subjected to regular, mass-mortality events (and potentially being more likely to attain densities close to K).

2.4.3 *r*- and *K*-selection

Robert MacArthur and Edward Wilson (1967), on the other hand, were very explicit in their treatment of density-dependent selection. They referred to organisms in uncrowded, low-density environments as being subjected to '*r*-selection', as opposed to those in crowded environments that are subjected to 'K-selection'. In coining these terms, MacArthur and Wilson (1967) leaned heavily on the logistic model of population growth (Chapter 1; Equation 1.14). In uncrowded environments, they appropriately defined fitness as r (or r_{max}). But in crowded environments, they unhelpfully defined fitness as K (or carrying capacity) (MacArthur and Wilson 1967: 149).

Eric Pianka (1970) explicitly assigned life-history correlates to r- and K-selected species. He classified r-selected species as those that mature early in life at a small size, produce many small offspring, and live a short life. In contrast, K-selected species delay maturity until they have attained a large size, produce few large offspring, and live a comparatively long life.

From a density perspective, the idea is as follows. When density (and intra-specific competition) is low, organisms should be favoured to invest greater resources into reproduction and so to produce many small offspring, each of which should have a reasonable opportunity to survive within the low-competition environment. By contrast, when conditions are crowded, competitive interactions are assumed to be intense. Thus, high-density conditions are predicted to favour a strategy for which resources are preferentially allocated to enhancing competitive ability and body maintenance rather than reproduction, a situation thought to favour life histories that result in larger body size and the production of fewer but larger offspring.

2.4.4   *r-K-ic?*

Issues pertaining to r- and K-selection were evident from the start, beginning with Pianka (1970) himself. He described MacArthur and Wilson's (1967) invocation of the 'much overused' logistic equation as 'unfortunate' (Pianka 1970: 592). Yet, if one is interested in how density affects selection, the classic model of density-dependent population growth would seem to be a logical point of departure. This is because natural selection is a process that acts on individuals within populations, and the logistic model is a within-population representation of density-dependent population dynamics. Pianka later clarified that his objective in his 1970 paper was to apply the terms r- and K-selection to describe how natural selection results in adaptive differences *among* species (Pianka 1979). His intent was to put together a set of correlated variables that he felt were indicative of boom-and-bust, opportunistic species as opposed to species that tend to inhabit relatively stable environments.

Despite its lasting influence (having been cited ~ 4 500 times), the scheme proposed by Pianka (1970) has some significant limitations (Andersen (2019) provides a recent critique, focusing on fishes). Although the life histories of some groups of organisms appear to fall along an r–K continuum of trait covariation (e.g. mammals; Stearns 1983), some do not (e.g. salmonid fish; Hutchings and Morris 1985). Contrary to MacArthur and Wilson's (1967) explicit contention—and Pianka's (1970) implicit assumption—that fitness can be measured as K, natural selection does not directly act on carrying capacity (which is a function of the environment). By definition, K-selection should result in increased carrying capacity. Yet, there is no good evidence that increased competitive ability—among mammals at least—is associated with higher K (Promislow and

Harvey 1990). Contrary to expectation, if density is regulated through resource limitation, competition might be expected to select for larger body size and, hence, lower carrying capacity (Promislow and Harvey 1990). This renders the term '*K*-selection'
confusing at best.

2.4.5   The ubiquity of *r*-selection

One overarching troubling point associated with the *r-K* selection scheme is that *all*
organisms are *r*-selected, insofar as natural selection is expected to favour the life history
that maximizes r_{max} in any particular environment. Focusing on r_{max} can be a useful to
way to think about which classifications are likely to be more informative than others.

   Think of a life history as a solution that natural selection has produced to solve the
problem of how to persist in a given environment. Often the solution results in the same
r_{max} but the solution—that is, the life-history trait combination—is very different.
Consider, for example, a terrestrial mammal (the North American bobcat, *Lynx rufus*)
and a marine fish (Atlantic cod, *Gadus morhua*). The r_{max} of bobcats is similar to that of
early-maturing, southern populations of cod (~0.9; Hutchings ct al. 2012). Both have
roughly similar maximum lifespans (32 yr for bobcats; 25 yr for cod) and similar ages at
maturity (bobcat: 1–2 yr; southern cod: 2–3 yr) (de Magalhães and Costa 2009;
Myhrvold et al. 2015; Myers et al. 1997). But these species differ considerably in terms
of offspring size and number. Bobcats produce 2–3 large (0.25 kg) young per breeding
cvent while cod produce millions of 1.5 mm diameter eggs.

   Comparing two marine fishes, the tiger shark (*Galeocerdo cuvier*) and cod inhabiting
the northwest Atlantic have similar values of r_{max} (~0.2–0.3) but the tiger shark produces
10–82 large offspring per spawning event as opposed to the cod's millions (Hutchings
et al. 2012). As with southern cod populations and bobcats, despite a 10^5–10^6 difference
in fecundity, northwest Atlantic cod and tiger sharks have simply evolved different life-
history solutions that have resulted in a similar value of r_{max}.

2.4.6   Fast-slow continuum

These comparisons suggest that there is merit in thinking about life-history trait covariation within the context of maximum per capita rates of increase. One way to do this is
by explicitly accounting for species differences in natural mortality, *M*. A key means by
which *M* is related to r_{max} is through body size. Increases in body size are associated with
declining *M* across multiple taxa (McCoy and Gillooley 2008; Figure 2.18). Body size
is also negatively associated with r_{max} (Fenchel 1974; Hutchings et al. 2012) (see
Chapter 9). These patterns of covariation are one means of linking *M* with r_{max}. Another
is through considerations of lifespan. The greater the rate of natural mortality, the shorter
the lifespan. Short lifespans are also associated with small body size (Figure 2.12). Thus,
the greater the value of *M*, the smaller the body size of the organism, the briefer the
lifespan, and the higher the r_{max}.

   A widely adopted life-history classification scheme is explicitly linked to species
differences in natural mortality. This is the 'fast-slow' continuum of life-history trait

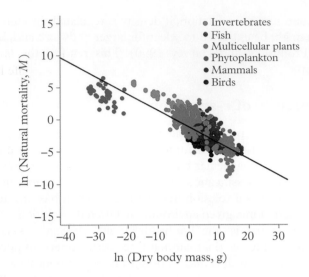

Figure 2.18 *Natural mortality (M; corrected for temperature) as a function of body mass for a broad selection of plants and animals. The line represents a least-squares regression. Data have been ln-transformed.*

From McCoy and Gillooly (2008). Reprinted by permission from John Wiley and Sons.

covariation, a concept that emerged from studies of mammal life histories in the early 1980s (Stearns 1983; Harvey and Clutton-Brock 1985). These studies identified body size as a key covariate of life histories. After accounting for species differences in body size, Promislow and Harvey (1990) found natural mortality to be the best predictor of variation in mammalian life-history traits.

Mammals with relatively high M tend to mature at a young age, produce many, comparatively small offspring, and live short lives. These are the characteristics of species with 'fast' life histories. Those with slow life histories experience low M, delay maturity, produce relatively few large offspring, and live long lives. Although species with fast and slow life histories tend to be comparatively small and large, respectively, the patterns of trait covariation are evident even if the effects of body size are removed from the analysis (Promislow and Harvey 1990). The fast-slow continuum has since been extended to species beyond mammals.

2.4.7   Pace-of-life syndrome

There have been attempts to build upon this continuum by incorporating potential metabolic, physiological, and behavioural correlates of fast and slow life histories. Ricklefs and Wikelski (2002), for example, drew attention to the role of physiology in fast-slow life histories, arguing that the small clutches and long lifespans of tropical passerine birds may be indicative of a 'syndrome of a slow pace of life' (Ricklefs and Wikelski 2002: 466). Wikelski et al. (2003) extended the pace-of-life concept to include metabolic rate, finding that the rate of metabolic turnover was lower among birds having slow life histories when compared with birds that have fast life histories. Indeed, the allocation of metabolic energy to traits that determine fitness and the pace of living figures prominently in recent arguments for a metabolic theory of life histories (Burger et al. 2019).

Réale et al. (2010) articulated the pace-of-life syndrome hypothesis, arguing that behavioural differences among individuals, i.e. their 'personalities', covary with life-history traits, physiology, and metabolic rate. They suggested that 'proactive' individuals (aggressive, bold, superficial explorers) should exhibit a fast pace of life, whereas 'reactive' individuals (lowly aggressive, shy, thorough explorers) should exhibit a slow pace of life. A relatively succinct way of expressing the pace-of-life syndrome hypothesis is to say that individuals with a slow pace of life are expected to grow slowly, delay reproduction, live longer lives, develop stronger immune responses, and avoid 'risky' situations when compared to individuals with a fast pace of life. Although the pace-of-life hypothesis continues to generate considerable interest, broad-scale analyses have not yet documented firm support for its underpinnings (e.g. Royauté et al. 2018).

2.5   Summing Up and a Look Ahead

There is extraordinary variability in species life histories. This is especially true across taxa that are phylogenetically diverse. But it can also be true among populations within a single species. Within the subphylum Vertebrata, this variation is strikingly evident from frequency distributions of several life-history traits, including age at maturity, fecundity, offspring size, and lifespan. Much of the variability in life-history traits can be attributed to variability in body size. Patterns of trait variability can reveal constraints. One type of constraint prevents organisms from expressing trait values commonly observed in other taxa, while a second type constrains covariation between traits, producing life-history invariants.

Patterns of covariation have suggested putatively adaptive groupings of life-history traits. For plants, one enduring classification scheme focuses on the influence of competition, disturbance, and stress, distinguishing strategies that differ in life history and habitat. Among animals, the early *r*- and *K*-selection continuum has given way to one that distinguishes fast from slow life histories, one variant of which has been termed the pace-of-life syndrome. Common to each life-history classification scheme is the presumptive ubiquity of selection, leading naturally to a discussion in Chapter 3 of the underlying roles of genes, genetic architecture, phenotypic plasticity, and reaction norms in life-history evolution.

3

Genetic Variation and Phenotypic Plasticity

3.1 Genetic Underpinnings

3.1.1 Distinguishing phenotype from genotype

Chapter 2 focused on phenotypic variation in life-history traits, i.e. variability that is physically observable. When measuring phenotypic variability, we are measuring an 'end result' or end product of underlying causal processes. Consider the height of a plant at maturity or the fecundity of a sea turtle. These are phenotypic (observable) traits that in both cases are influenced by genes, the environment, and interactions between genes and the environment.

Put simply, phenotype (P) is a function of genotype (G), environment (E), and genotype-by-environment interactions $(G \times E)$.

$$P = G + E + G \times E \qquad \qquad \text{Equation 3.1}$$

The initial part of this chapter focuses on the genotype (G). When considering a specific trait x, the genotype can be thought of as that part of the genetic make-up of an organism that determines, or is causally responsible for, trait x. Using the concept of genetic architecture, we explore how the underlying genetic basis of a trait can be caused by many genes, each having a small phenotypic effect, and/or few genes each having a large effect. There are also situations in which traits are influenced by groups of many genes linked together on the same chromosome that are inherited as single units or 'supergenes'.

We then examine how a particular type of genetic variability (additive genetic variance) is related to trait heritability (i.e. the similarity in trait phenotype between parents and offspring), and how heritability helps to determine how rapidly a trait will change in response to natural or human-induced selection.

Thereafter, following a section on how the environment can affect the phenotypically plastic expression of traits, we explore genotype-by-environment $(G \times E)$ interactions, using reaction norms. These are visually heuristic and intuitively tractable depictions of how a trait varies with an environmental factor (the most common type of reaction

A Primer of Life Histories: Ecology, Evolution, and Application. Jeffrey A. Hutchings, Oxford University Press. © Jeffrey A. Hutchings 2021.
DOI: 10.1093/oso/9780198839873.003.0003

norm) or with another trait (less common). Reaction norms have long been used to study phenotypic plasticity. Today, they are increasingly seen as an invaluable tool for examining genetic differences in how individuals and populations respond to environmental change, such as global warming.

3.1.2   Genes and phenotypes

The word 'gene' can mean many things. Strictly speaking, a gene comprises a sequence of nucleotides in DNA (deoxyribonucleic acid) and/or RNA (ribonucleic acid). Genes provide the genetic code for the synthesis of gene products, such as proteins, that contribute to the phenotypic expression of a given characteristic (such as age or size at maturity). Genes differ among individuals. Gene variants are called alleles. Most allelic differences have no discernible influence on the phenotype (termed 'neutral' mutations), whereas others do.

As noted briefly in Chapter 1, Mendel's work on pea (*Pisum sativum*) hybridization laid the foundation for linking phenotypic variability with genetic variability. One of his experiments involved flower colour. Some plants produce purple flowers; others white (Figure 3.1). The colour purple is produced by anthocyanin pigments. It turns out that the production of anthocyanin pigment is regulated by a set of multiple genes. This in itself is not surprising. What is interesting is that the protein responsible for turning this set of multiple genes on or off is regulated by just one gene (Hellens et al. 2010).

In fact, it is simpler than that. This regulatory 'switch' that turns the pigment-producing genes on is determined by a single nucleotide within this single gene. If the nucleotide is guanine, the gene makes the protein responsible for expressing the pigment-producing genes, and the flower colour is purple. But if this nucleotide is adenine, the regulatory gene is prevented from making the protein, the set of pigment-producing genes remains unexpressed, and the flower colour is white.

Figure 3.1 *White and purple flowers of the pea,* Pisum sativum. *Photo attributions: white flower © Dyorkey CC BY-SA 3.0; purple flower © Forest and Kim Starr CC BY 3.0. Both photos were cropped at their edges.*

Genetic control of flower colour in these peas illustrates both single- and multi-locus control on the expression of a trait. The production of the anthocyanin pigment (which determines flower colour) is determined by many genes each having a correspondingly small effect. This is termed 'multi-locus' or 'polygenic' control. (The location of a gene within a genome is called its 'locus'.) However, the ultimate genetic determinant of whether a flower is white or purple is genetically controlled by a single regulatory gene, at a single locus, with a large effect (insofar as a change in colour can be termed a large effect). If the regulatory gene is expressed (switched on), the flower is purple.

The question of whether organismal characteristics are generally controlled by many independent genes each having small effect (the multi-locus model) or single genes with large effect (the single-locus model) has a long history, dating back to the early twentieth century. Long before his fundamental work on life-history evolution (sub-section 1.1.3), Fisher explored the general question of how genetic variation conforming to the Mendelian pattern of inheritance contributes to phenotypic variation. He concluded that if many genes are responsible for affecting the phenotypic value of a single trait, then a random sampling of alleles at each gene will produce a continuous, normally distributed set of phenotypes within a population (Fisher 1918). In other words, an underlying genetic architecture of multi-locus control of a phenotype is sufficient to produce the normal, continuous distributions of phenotypic traits often observed in wild populations.

Given that we regularly observe phenotypic traits, such as body size at maturity, to have normal (or nearly normal) distributions within populations, it has long been concluded, based on the work of Fisher (1918) and others, that such traits are controlled by the cumulative actions of numerous genes each of which has a small effect on the ultimate expression of the phenotype. These traits are termed quantitative or continuous traits because they can be measured in quantitative (numerical) terms and because they tend to have comparatively smooth, continuous distributions. (In contrast, flower colour in the peas mentioned above is termed a discontinuous or discrete trait.) The nature of the genetic architecture underlying a trait can influence how that trait responds to selection, one example of which will be discussed in sub-section 3.2.4.

3.1.3   Quantitative genetics and partitioning genetic variance

Life-history traits are generally considered to be quantitative traits and they are almost always modelled as such (Lande 1982). Their multi-locus form of genetic architecture lies at the heart of the research field of quantitative genetics. A quantitative trait is a trait whose value varies continuously among individuals within a single population. These differences among individuals can be described by the variation in trait values. A fundamental objective of quantitative genetics is to partition the genetic variation (V_G) underlying the phenotypic variation (V_P) into different sources (Falconer and MacKay 1996).

Recall Equation 3.1, which describes the relationship between the phenotypic value of a trait, the genotype underlying that trait, and the environment, i.e. $P = G + E + G \times E$. Moving from the individual level to the population level, the partitioning of trait variance can be illustrated simply by modifying this equation such that one is able to express

V_P as a function of V_G, environmental variance (V_E), and the variance associated with gene × environment interactions ($V_{G \times E}$), such that:

$$V_P = V_G + V_E + V_{G \times E} \qquad \text{Equation 3.2}$$

The genetic variance, V_G, can be broken down further:

$$V_G = V_A + V_D + V_I, \qquad \text{Equation 3.3}$$

into additive variance (V_A), dominance variance (V_D), and epistatic variance (V_I).

If you are interested in understanding the basics of selection within the context of life-history evolution, it is the additive genetic variance (V_A) that is of greatest interest because trait heritability is determined by V_A. Additive variance pertains to independent, additive effects of all the alleles that affect the phenotypic variation of a specific trait. These independent effects sum to the character value of the trait. If one allele causes a phenotypic shift in the value of a trait (relative to the population mean) by an amount x_1, and a second allele causes it to differ by an amount x_2, the total difference in the trait value from the population mean for individuals bearing both genes is $x_1 + x_2$. The greater the additive variance in a trait, the greater that trait's heritability.

To illustrate additive variance, consider a hypothetical example (Figure 3.2). Assume that the length of wings in an insect is determined by a single locus that has two alleles: the homozygous $W_1 W_1$ genotype produces wings that are long (10 mm) whereas the homozygous $W_2 W_2$ genotype produces wings that are short (4 mm). If all the genetic variance in wing length is additive, both alleles are expressed to equal degrees and the $W_1 W_2$ genotype (and $W_2 W_1$ genotype) produces wings that are intermediate in length (7 mm).

In contrast to these additive effects, V_D and V_I refer to those parts of the genetic variance caused by interactions between alleles. Dominance variance (V_D) originates from interactions between alternative alleles at the same locus. In the example illustrated by Figure 3.2, if the expression of alleles W_1 and W_2 is not equal, such that one allele is

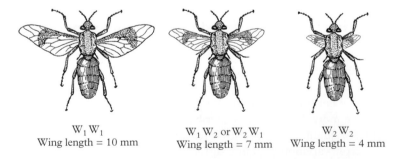

$W_1 W_1$
Wing length = 10 mm

$W_1 W_2$ or $W_2 W_1$
Wing length = 7 mm

$W_2 W_2$
Wing length = 4 mm

Figure 3.2 *Wing length controlled by two alleles at a single locus. Each W_1 allele contributes 5 mm to the length of a wing; each W_2 allele contributes 2 mm to the length of a wing. If all of the genetic variance in wing length is additive, the actual wing length will be equal to the sum of the length contributions of each allele.*

dominant over the other, the W_1W_2 genotype would produce a wing that was relatively long if the W_1 allele was dominant or a wing that was relatively short if the W_2 allele was dominant. Epistatic variance (V_I) originates from interactions between alleles at different loci, such that alleles act differently depending on what other alleles are present. (The pea flower-colour example described above in sub-section 3.1.2 provides an example of epistatic variance.)

3.2 Measuring Evolutionary Change in Response to Selection

3.2.1 Heritability and additive genetic variance

As known from basic genetic principles, each diploid parent passes on a single allele per locus (one allele at each locus, one locus on each chromosome) to each of its offspring, meaning that parent-offspring resemblance depends on the average or additive effect of single alleles. Thus, it is the additive component of the overall genetic variance (V_A) that we are interested in when examining trait heritability, i.e. the resemblance between parents and offspring.

Heritability (h^2) is the proportion of the total phenotypic variance in a trait attributable to additive genetic variance, such that:

$$h^2 - V_A/V_P \hspace{4cm} \text{Equation 3.4}$$

h^2 ranges between 0 (no resemblance between parents and offspring) and 1 (parents and offspring are identical in their expression of a trait). Heritability can be estimated from a controlled breeding programme that allows one to estimate V_A separately from V_D and V_P, an approach that is quite labour-intensive and time-consuming. An alternative approach is to construct a parent-offspring regression by plotting the average trait value of offspring against the mid-parent values for the same trait: the slope of such a regression is equal to h^2.

3.2.2 Response to selection

One of the primary causes of genetic or evolutionary change is selection. Selection results from situations in which the mean phenotype of individuals that successfully breed (those that are 'selected') differs from the mean phenotype of a random sample of individuals in the same population. Selection has three conditions. The first is that phenotypes must be variable; if there is no phenotypic variability, there is nothing to 'select'. Second, there must be a link between phenotypic variability and reproductive success, meaning that some phenotypes have greater reproductive success than other phenotypes. Third, in addition to phenotypic variability and differential reproductive success, the phenotypic trait(s) must be heritable (i.e. h^2 must be significantly greater than 0).

These three conditions are met in life-history traits. As illustrated in Chapter 2, they exhibit high degrees of phenotypic variability. And as the example of Atlantic salmon life-history variation in Chapter 2 implies (female length and fecundity at maturity ranging between 1.0 to 2.5 orders of magnitude; Figure 2.8), reproductive success can differ substantially among life-history phenotypes. In terms of heritability, mean h^2 estimates for life-history traits have consistently been shown to be significantly greater than zero: (i) 0.26 for 75 animal species (excluding *Drosophila* spp. and humans; Mousseau and Roff 1987); (ii) 0.12 for *Drosophila* spp. (Roff and Mousseau 1987); and (iii) 0.11 and 0.02 for age at birth of first child and number of children, respectively, in humans (Gavrus-Ion et al. 2017).

Selection can be generated with and without human intervention. The latter refers to natural selection and sexual selection. The former can either be intentional, such as the deliberative breeding of plants or animals to generate characteristics desirable to humans (Figure 3.3), or unintentional, resulting from factors such as habitat alteration (e.g. construction of barriers to migration), introduction of non-native species, or human-induced alterations of ecosystems (one result of over-exploitation; see Chapter 10).

Once the conditions for selection have been established (phenotypic variation, differential reproductive success, trait heritability), we can then ask whether the magnitude of the response to selection is likely to be marginal or considerable. To address this question, researchers often use the 'breeder's equation', or a variant thereof, which describes

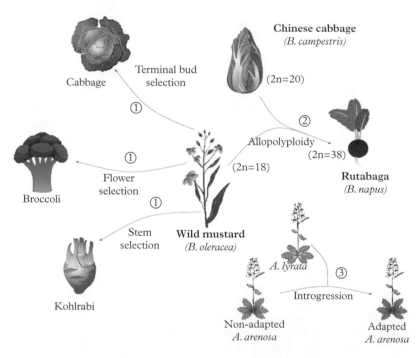

Figure 3.3 *An example of artificial selection. Common vegetables that were cultivated from forms of wild mustard,* Brassica oleracea. *This is evolution through artificial selection.*

Source: Fernie and Yan (2019). Reprinted by permission from Elsevier.

how the response of a single trait to selection (*R*) can be predicted with knowledge of the trait's heritability (*h²*) and what is termed the selection differential (*S*), such that:

$$R = h^2 S$$
Equation 3.5

S is the difference in the mean phenotypic value of the trait between the selected population and the population if breeding occurred at random (Figure 3.4).

The origin of the breeder's equation is frequently attributed to Lush (1937), an animal geneticist whose book *Animal breeding plans* laid the foundation for using trait heritability and the selection differential to predict selection responses in animal breeding programmes (Ollivier 2008). Although its origins lie in agricultural genetics, the breeder's equation has been seminal in the application of quantitative genetics to predicting rates and directions of evolutionary change in life histories.

3.2.3   Genetic trade-offs

The breeder's equation exudes simplicity when understanding the evolution of single traits. It is highly influential in studies of genetic change brought about by natural and artificial selection. It is perhaps most useful in predicting the rate and magnitude of change during a short-term (<3–5 generations), 'major' shift in the strength of selection,

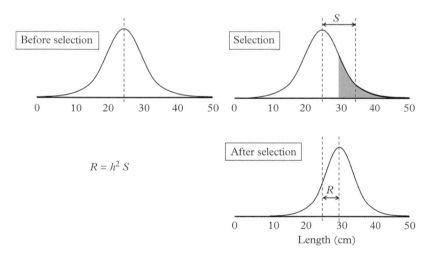

Figure 3.4 *The response to selection of a life-history trait (R) is a function of the heritability of the trait (h²) and the selection differential (S). In this example, before selection, the mean length at maturity in the population is 25 cm. Selection, acting against individuals maturing smaller than 30 cm, results in a selected population of breeders (grey shading) averaging 35 cm at maturity (S = 35 − 25 = 10 cm). If h² = 0.5, the average length at maturity in the next generation will be 5 cm (R = 10 cm × 0.5) longer than it was in the previous generation. Thus, in this example, one generation of rather strong selection on a trait with moderately high heritability has increased the average length at maturity from 25 to 30 cm.*

such as from weak to strong selection (e.g. an environmental change that suddenly favours small, narrow beaks in Darwin's finches, *Geospiza* spp., when formerly large, deep beaks were favoured; Grant 1986).

Longer-term predictions regarding selection responses can be problematic (Barton and Turelli 1986). This is because traits do not exist in isolation of one another. The evolution of a single trait rarely occurs independently of the evolution of other traits. Consider the example of a negative genetic correlation existing between two traits. Such correlations can arise because single genes can affect each trait in different ways (Figure 3.5). This is called 'pleiotropy'; pleiotropy occurs when a single gene influences two or more traits. A gene might have a positive effect on one trait, such as size at maturity, but also have a negative effect on another trait, such as size of offspring. If multiple genes have similar pleiotropic consequences for these two traits, a negative genetic correlation between the two traits will result (Figure 3.5).

What are the consequences of negative genetic correlations from a selection perspective? If a negative genetic correlation exists between two traits, selection favouring an increase in one of those traits will result in 'correlational selection' that produces a decrease in the other trait. Using the example in Figure 3.5, selection favouring increased size at maturity will also result in a reduction in the size of offspring. While larger size at maturity might be advantageous, resulting in an increase in fitness, smaller size of offspring might not be advantageous, resulting in a decrease in fitness. This is termed a genetic 'trade-off'. Genetic trade-offs can, thus, affect the magnitude of a selection response.

3.2.4   Genetic architecture

Genetic architecture refers to how a phenotypic trait is controlled by one or more genes. It accounts for interactions among alleles (e.g. number and effect sizes of contributing loci, dominance, epistasis), structural arrangements of genes on a chromosome, and the degree to which different loci are linked or associated with one another (Oomen et al. 2020).

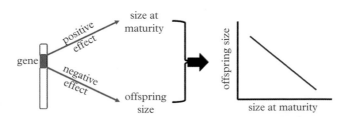

Figure 3.5 *Pleiotropy exists when a single gene influences two or more phenotypic traits. Here, a single gene has a positive effect on size at maturity but a negative effect on offspring size. When multiple genes have similar consequences for these two traits, a negative genetic correlation can arise between them (red line), such that selection for increased size at maturity is correlated with decreased size of offspring.*

A key question is whether genetic control of a phenotypic trait is typically the product of few genes with large effect or multiple genes each having small effect. It is generally assumed that life-history traits are controlled by many genes of small effect (Stearns 1992; Roff 1992, 2002). However, the rapid development of genomic technology has revealed an increasing number of large-effect loci and 'supergenes' (Rieseberg 2001; Oomen et al. 2020).

The Atlantic salmon again provides an informative example. In 2015, it was reported that a single gene (hypothesized to be *vgll3*, a transcriptional co-factor gene) is responsible for almost 40 per cent of the variation in age at maturity in this species (Barson et al. 2015). This surprising finding ran counter to expectations that this life-history trait is controlled by multiple, small-effect genes. Mathematical model simulations were run to examine how single- versus multi-locus genetic architecture might affect selection responses to size-selective mortality.

Under the multi-locus model, selection against older salmon favoured genotypes that matured at younger ages (Figure 3.6, upper panel), resulting in unidirectional evolution towards earlier age at maturity (Kuparinen and Hutchings 2017). However, under the assumption of single-locus control, similar selection pressure did not produce clear phenotypic trends (Figure 3.6, lower panel). This appears to be the case with and without sexually dimorphic expression (Oomen et al. 2020).

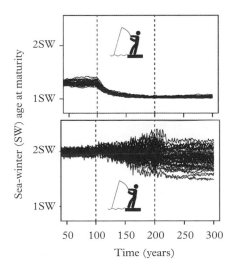

Figure 3.6 *Simulated evolutionary dynamics of age at maturity (winters at sea) in Atlantic salmon exposed to selective fishing. Dashed vertical lines identify periods (left to right) of (i) no fishing; (ii) removal of 36% of the population annually by fishing; and (iii) recovery during which fishing had ceased. The upper panel shows the results for 50 simulated populations (each line represents a simulation) for a genetic architecture of multi-locus control of age at maturity; fishing favours earlier maturity. In the lower panel, which shows the results under single-locus control, age at maturity does not exhibit a discernable trend.*

Source: Kuparinen and Hutchings (2017).

3.3   Phenotypic Plasticity

Chapter 3 opened with a central tenet of evolutionary biology: phenotype (P) is a function of genotype (G), the environment (E), and ways in which genes and the environment interact $(G \times E)$. This last term—gene-by-environment $(G \times E)$ interactions—can be reflected by phenotypic plasticity: the ability of a genotype to produce different phenotypes across an environmental gradient. All humans, for example, respond to increased relative humidity (environmental gradient) by increasing their rate of perspiration (phenotypic response). That is plasticity. But we also know from personal experience that as relative humidity increases, some people (i.e. genotypes) respond by perspiring to greater degrees than others. This is an example of a $G \times E$ interaction; different genotypes responding to the same environmental gradient but in different ways.

The history of research on phenotypic plasticity is a rich one, beginning perhaps with Baldwin's 'new factor in evolution': individuals differ not only in phenotype but also in the way that the phenotype can be altered by changing environmental circumstances (Baldwin 1896). Studies of phenotypic plasticity have tended to focus, albeit not exclusively, on ectotherms (Table 3.1). These include most plants, fishes, amphibians, reptiles,

Table 3.1 *Selected examples of how environmental variables affect the expression of life-history traits.*

Species	Environmental variable	Life-history traits affected	Citation
African butterfly (*Bicyclus anynana*)	temperature; humidity	lifespan; age at maturity	Brakefield and Zwaan (2011)
Multiple species, including fruitfly (*Drosophila melanogaster*); primates	diet	lifespan	Flatt (2014) and references therein
Red deer (*Cervus elpahus*)	temperature	offspring size	Nussey et al. (2005)
Many plants, such as ryegrass (*Lolium perenne*) and creeping bentgrass (*Agrostis stolonifera*)	temperature; nutrients	leaf growth; root growth	Schlichting and Pigliucci (1998)
Yarrow (*Achillea lanulosa*)	altitude	height	Clausen et al. (1948)
Lady's thumb (*Polygonum persicaria*)	soil moisture	fruit biomass	Sultan and Bazzaz (1993)
Fishes, multiple species	temperature, salinity, oxygen, pH, food supply	age at maturity; growth; offspring size; offspring survival	Hutchings (2011); Oomen and Hutchings (2015)
Narrowleaf hawksbeard (*Crepis tectorum*)	light	seed length	Andersson and Shaw (1994)

and insects, although some species in each of these categories have a limited ability to thermoregulate some tissues or organs (even plants; Michaletz et al. 2015).

Plants have long provided some of the most striking examples of phenotypic plasticity (as detailed by Schlichting and Pigliucci 1998). In his influential book entitled *Factors of evolution: the theory of stabilizing selection*, Schmalhausen (1949) discussed the tremendous differences in leaf development and shape exhibited by arrowheads (*Sagittaria sagittifolia*), depending on whether the plant grows in a terrestrial environment or is submerged by water. This phenomenon is known to exist in other species, such as white water-crowfoot (*Ranunculus aquatilis*) (Figure 3.7) and yellow water buttercup (*R. flabellaris*).

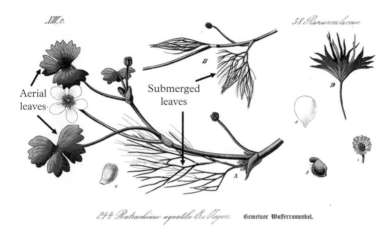

Figure 3.7 *Phenotypic plasticity in leaf shape in the white water-crowfoot* (Ranunculus aquatilis). *On the left are the flat, lobed leaves produced when the plant grows in a terrestrial environment. In the centre are the needle-like leaves produced by the same plant when it is submerged.*

Original book source: Prof. Dr Otto Wilhelm Thomé Flora von Deutschland, Österreich und der Schweiz *1885, Gera, Germany.*

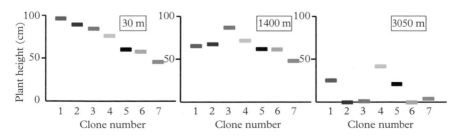

Figure 3.8 *Differences in mean size at maturity (height, in cm) among seven clones of yarrow* (Achillea lanulosa) *at three different altitudes: 30 m; 1400 m; 3050 m.*

Data source: Clausen et al. (1948).

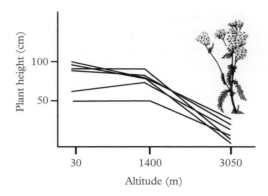

Figure 3.9 *Reaction norms for plant height among seven clones of yarrow across altitude. Inset: water colour of* Achillea lanulosa © *Mary Vaux Walcott.*

A classic and more instructive example from a $G \times E$ perspective was provided by Clausen and colleagues who examined how seven different clones of yarrow (*Achillea lanulosa*) grew at three different altitudes in California (Clausen et al. 1948). One can readily see that there are clonal, i.e. genetic, differences in plant height at different altitudes (Figure 3.8).

Most importantly, in terms of $G \times E$ interactions, the rank order of clonal height changes with altitude. Clone 4, for example, is the fourth highest at 30 m, second highest at 1400 m, and tallest at 3050 m. $G \times E$ interactions become visually evident in Figure 3.9, which illustrates how plant height changes with altitude for each of the seven clones. Each of the seven lines in this figure is a 'reaction norm', a graphical representation of how a genotype varies its phenotype across an environmental gradient.

3.4   Norms of Reaction

Norms of reaction (e.g. Figure 3.9) are linear or non-linear functions that characterize the pattern with which the phenotypic value of a trait, for a given genetic entity, changes with the environment (Oomen and Hutchings (2020) compiled a bibliography on reaction norms). The German zoologist Woltereck (1909) introduced the term *Reaktionsnorm* to describe clonal variation in how the crustaceans *Daphnia* spp. and *Hyalodaphnia* spp. alter the height of their head (relative to the length of their body) as a function of the amount of available food (algae). The importance of Woltereck's work was recognized immediately. Johannsen (1911) (who coined the word 'genotype') concluded that *Reaktionsnorm* were 'fully compatible with the genotype-conception' (Johannsen 1911: 990) and were, thus, of potential importance in evolution.

In the 1930s, Dobzhansky, one of the great thinkers of evolutionary biology, gave voice to what he perceived to be an overlooked yet fundamentally important element of

evolutionary change, namely the ways in which mutation and selection act upon heritable variability in a genotype's norm of reaction (Dobzhansky 1937: 169):

> One must constantly keep in mind the elementary consideration which is all too frequently
> lost sight of in the writings of some biologists; what is inherited in a living being is not this
> or that morphological character, but a definite norm of reaction to environmental stimuli.
> . . . [A] mutation changes the norm of reaction.

Fully consistent with Dobzhansky's prescience, there is now abundant evidence that
reaction norms can be genetically variable and respond to selection (Oomen and
Hutchings 2020).

    The basic elements of reaction norms are straight forward. It can be helpful to think
of three general patterns. In the first, all reaction norms share the same slope—zero—but
differ in intercept (Figure 3.10, upper panel). In this case, there is no plasticity because
the trait does not change with changes to the environment. In the middle panel of
Figure 3.10, the reaction norms share the same non-zero slope. We can interpret this as
meaning that each genotype (or family or population, whatever our unit of measure happens to be) is plastic (the trait varies with the environment), but that there are no $G \times E$
interactions in plasticity (the plastic responses are the same among genotypes). In the
bottom panel of Figure 3.10, we have reaction norms that differ in slope (and intercept).
Crossing reaction norms, reflecting different reaction-norm slopes, are indicative of $G \times E$
interactions; genotypes differ in how they respond plastically to environmental change.

    Crossing reaction norms illustrate the utility, arguably the necessity, of varying the
environment when undertaking common-garden experiments. For example, imagine

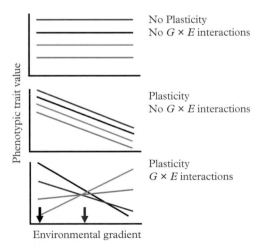

Figure 3.10 *Reaction norms, illustrating different levels of plasticity and genotype-by-environment
(G × E) interactions. The red arrow represents an environment in which reaction norms cross and the
genotypes express similar phenotypes; the black arrow represents an environment in which the genotypes
express dissimilar phenotypes.*

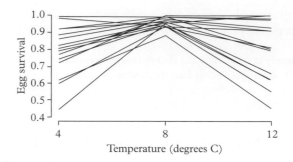

Figure 3.11 *Reaction norms illustrating how the relationship between egg survival and temperature in chum salmon differs among families (each reaction norm represents a different family). Similar phenotypes (i.e. survival) are expressed among families at 8° but not at 4° or 12°.*

Based on data from Beacham and Murray (1985).

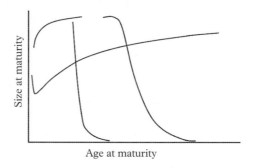

Figure 3.12 *Hypothesized bivariate reaction norms between age and size at maturity whose shape and position depend on relationships between growth rate and mortality. More details are provided in Figure 2.10.*

Reprinted by permission from John Wiley and Sons.

rearing different genotypes at a single temperature that happens to be very close to where the four crossing reaction norms intersect (red arrow in Figure 3.10). We might well conclude that the genotypes (or populations) did not differ from one another because their phenotypes would be quite similar. But if this imaginary experiment had included a second temperature (black arrow) that happened to be an environment in which the phenotypes differed considerably (as reflected by the reaction norms we are trying to uncover), we would conclude that the genotypes differed in their plastic responses to temperature change. An empirical example of this is offered in Figure 3.11, based on data on chum salmon, *Oncorhynchus keta* (Beacham and Murray 1985).

The concept of reaction norms has been used to describe associations between life-history traits. Early examples of such bivariate reaction norms were constructed by Stearns and Koella (1986) (Figure 3.12). (These were originally presented in Figure 2.10.) They hypothesized that individual growth and mortality were primary determinants of shape variability in bivariate maturation reaction norms for age and size at maturity. Their key point was that most organisms should mature neither at a fixed size nor at a fixed age, but along an age-size trajectory. By accounting for changes in growth and mortality, Stearns and Koella (1986) hypothesized that temporal changes in the shape and position of bivariate reaction norms could reflect evolved responses to selection. By the early 2000s, the development of probabilistic maturation reaction norms by Heino et al. (2002) opened up the possibility that Stearns and Koella's (1986) approach might be useful in disentangling growth-related phenotypic plasticity from genetic responses to fisheries-induced evolution (see Chapter 10).

It can be tempting to characterize differences in slopes, intercepts, or overall shapes of reaction norms as being adaptive (i.e. reflecting evolutionary processes that enhance fitness). But they need not be. The reaction norm, in whole or in part, might instead reflect physiological stress or a developmental constraint, rather than adaptation. This might especially be true when reaction norms are based on environmental values, such as excessively high or low temperatures, that fall outside of the range of values that organisms typically, or are reasonably likely, to experience (Figure 3.13).

One additional point to note is that the unit of study in reaction-norm research often differs among organisms. Studies at the genotypic level are not uncommon in many plants, clonal organisms, and a few model species, such as *Drosophila* spp., because of the relative ease with which single genotypes can be generated and their responses to environmental change documented. However, for most sexually reproducing organisms, the level of family (comprised of full-sibs) is the genetically finest at which reaction norms can be studied effectively. Another level commonly examined is that of populations, particularly when one is interested in whether populations of the same species are likely to differ genetically in the ways in which they respond to environmental change.

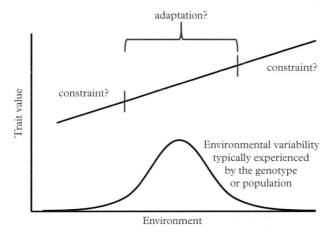

Figure 3.13 *The adaptive significance of a reaction norm likely depends on the degree to which the environmental gradient encompasses values that fall within and without the values typically experienced by an organism or population.*

3.5   Summing Up and a Look Ahead

The phenotypic expression (P) of a life-history trait is a function of an individual's genotype (G), environment (E), and ways in which the genotype interacts with the environment ($G \times E$ interactions). The underlying genetic variability of a trait is the result of additive effects among alleles at multiple genes (loci) (V_A) and non-additive effects caused by allelic interactions at the same locus (V_D) or at different loci (V_I). Additive genetic variance determines the resemblance of a trait (the heritability) between parents and offspring; it is a primary determinant of the response by organisms to natural and human-induced selection. The response of a trait to selection also depends on the underlying genetic architecture. Genotype-by-environment ($G \times E$) interactions can be reflected by phenotypic plasticity which can be made visually and analytically tractable by reaction norms.

One of the key points to emerge from this chapter is that genes do not act in isolation of one another. This is one of the reasons why it can be quite challenging to reliably predict evolutionary responses to selection. What is true of the genotype is also true of the phenotype.

To greater or lesser degrees, life-history traits are usually correlated with one another. Many of these correlations are negative, which can be indicative of a trade-off insofar as an increase in one trait is traded-off against a decrease in another trait. As highlighted next in Chapter 4, few trade-offs in life-history evolution are as important as that between the effort expended towards reproduction and the cost realized by that effort.

4

Reproductive Effort and Costs

4.1 Trading Off One Set of Fitness Benefits for Another

Integral to the study of life-history evolution is the concept of a trade-off. Benefits derived from making one life-history 'decision' are made at a cost of not realizing potential benefits associated with alternative decisions. Consider, for example, the trade-offs associated with maturing (i.e. reproducing for the first time) early in life at a comparatively young age, as opposed to later in life at an older age (Figure 4.1). One clear benefit of early maturity is an increased probability of surviving to reproduce. The longer an organism delays maturity, the greater its chance of dying before it can reproduce. A second consequence of earlier maturity is shorter generation time. The shorter the generation

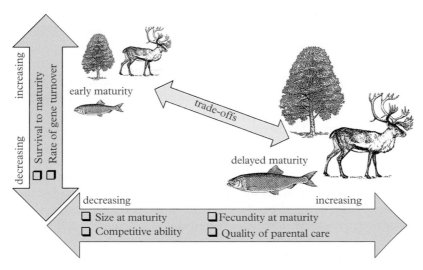

Figure 4.1 *Trade-offs associated with early versus delayed maturity. Maturing early in life carries with it the benefits of increased probability of surviving to reproduce and increased rate at which genes are 'turned over' or represented in a population. These benefits are traded-off against benefits of delaying maturity: larger body size; increased competitive ability; higher fecundity at maturity; better quality of parental care.*

A Primer of Life Histories: Ecology, Evolution, and Application. Jeffrey A. Hutchings, Oxford University Press. © Jeffrey A. Hutchings 2021.
DOI: 10.1093/oso/9780198839873.003.0004

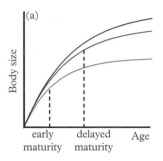
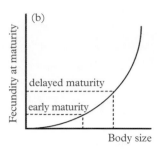

Figure 4.2 *Fecundity costs of reproduction. (a) Early (green) and late (red) maturing individuals experience slower growth, leading to smaller sizes at later ages, than non-reproductive individuals (black) because of the diversion of energy from growth to reproduction. (b) The smaller body size of earlier-maturing individuals results in lower fecundity at maturity compared to individuals that delay maturity. The black curve reflects the observation that fecundity increases curvilinearly with body mass in many indeterminately growing organisms.*

time, the more rapidly a genotype is able to represent, or 'turn over', its genes relative to other genotypes in the same population. The greater the representation of one's genes relative to those of other genotypes, the higher the fitness.

However, as beneficial as early maturity can be from a fitness perspective, it can come at multiple costs, especially for indeterminately growing organisms. A plant or fish or reptile that matures at its earliest opportunity will mature at a relatively small size and grow more slowly at subsequent ages than those that delay maturity. As a consequence, they will not attain as large a size at later ages when compared to those that have delayed maturity. The reason for this is that maturity requires that an organism divert its internal resources from growth (and/or survival and body maintenance) to reproduction (see section 4.2). Reduced growth produces an associated reduction in future fecundity because of the strong relationship that exists between seed/egg number per individual and reproductive size in most plants, fish, and reptiles (Figure 4.2). This re-allocation of energy can be interpreted as a 'fecundity or growth' cost to early maturity, insofar as the early-maturing organism is less likely to reap benefits that come with larger size.

For plants and animals, larger size can also be associated with enhanced competitive ability. Larger plants produce more extensive roots, giving them enhanced access to water and soil nutrients; the increased foliage of taller plants diminishes the light available to nearby plants competing for solar energy. Larger plants produce greater numbers of seeds; longer, heavier fishes produce greater numbers of eggs. In many animals, increased body size is associated with increased ability to acquire and defend territories and mates.

By delaying maturity, individuals might also be able to enhance the survival probability of their offspring. In some fishes, larger individuals produce larger, better-provisioned eggs; in many vertebrates, larger size allows for individuals to provide a higher quality of parental care than smaller individuals.

Of course, not all benefits are equal in terms of their positive contributions to fitness. For some populations, the increased likelihood of surviving to reproduce might be of greater value to the fitness of an early-maturing individual than the fecundity benefits of delaying maturity. Neither are all costs equally injurious. What matters is the strength or the magnitude of the trade-off in question. In a population subjected to selective exploitation of larger individuals, the probability of realizing a large body size, and reaping the associated benefit of increased fecundity, might be very low, thus magnifying the benefits of maturing at a young age. By contrast, in an unexploited population, for which the survival probability of attaining a comparatively old age is high, selection might well favour individuals that postpone reproduction. What differs between the fished and unfished populations is the nature of the trade-offs and how these relate to fitness.

4.2   The Nature of Trade-Offs

Trade-offs are the inevitable product of a constraint which prevents multiple positive outcomes from being simultaneously realized. Chapter 3 highlighted one such constraint—genetic architecture. Trait evolution rarely occurs independently of the evolution of other traits. One consequence of this constraint can be a negative genetic correlation between two traits that prevents selection from concomitantly increasing the value of both traits (Figure 3.5). This is an example of a genetic trade-off.

A second arguably more pervasive constraint results from an individual's allocation of fixed resources (Figure 4.3). Each organism, through various means (absorption of light

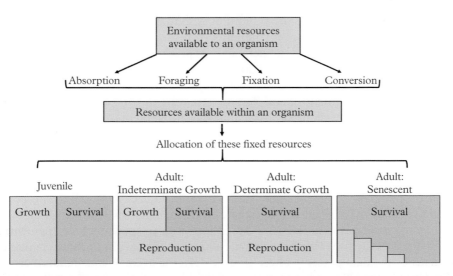

Figure 4.3 *Trade-offs can result from the allocation of fixed resources. Depending on developmental stage, resources are allocated to two or three primary purposes: growth, survival (including processes involved in body maintenance, without which the individual would die), and reproduction. Increased allocation of energy to one of these purposes would necessarily result in a reduction in the energy available to at least one other purpose. Senescent individuals can be typified by decreased allocation of resources to reproduction with increasing age.*

by plants, foraging by animals, fixation or conversion of chemicals by microbes), obtains resources from its external environment. These are stored, used for metabolic and other energy-demanding processes, converted into tissue, and excess or waste products excreted. The key premise from the perspective of trade-offs is that the resources available to an organism at any given time are finite or fixed, meaning that an increased allocation of energy to one purpose (growth, survival, reproduction) comes at the expense of allocating less energy for another purpose.

The nature of allocation-based trade-offs can be expected to change through an organism's life (Figure 4.3). Before maturity, resources (oils, lipids, proteins, and other nutrients) are allocated to physiological, hormonal, developmental, and behavioural processes or activities that promote either individual growth or survival (the latter term includes energy allocated to organismal maintenance and repair). After this juvenile stage, the indeterminately growing organism faces potential trade-offs between reproduction, survival, and growth whereas the determinately growing adult trades off resources between reproduction and survival. The senescent adult would not usually need to trade-off resources allocated to survival, although there are some mammals, insects, and birds that, beyond their reproductive ages, allocate energy to assist the rearing of related offspring through processes such as cooperative breeding and parental care (Lee 2003).

4.3   Reproductive Effort

Reproduction requires effort; effort requires resources. Reproductive effort can thus be defined as the proportion of total energy or resources allocated to all elements of reproduction. What constitutes an element of reproduction? It depends on the organism. For many angiosperms, reproductive effort includes not simply the energy required to produce gametes but resources allocated to pollinator-attraction mechanisms, fruit production, and various structures of the flower. For animals, in addition to the development of eggs and sperm, elements of reproduction might include migration to breeding grounds, mate competition, mate attraction (e.g. behavioural displays, vocalization, changes in colour or morphology), nest construction, ornaments, and parental care.

But while reproductive effort is easy to conceptualize, it is difficult to quantify its constituent parts, especially for organisms in the wild. One exception is the proportional allocation of body mass, or energy, to gamete production. In fishes, this has been approximated by the mass of the gonads (measured just prior to spawning) as a percentage of the total mass of the fish. This proportional allocation is typically higher in females than in males. For the data presented here (Figure 4.4), the average and standard error (se) for females is 12.6 ± 0.6 per cent, more than double that for males (4.8 ± 0.5 per cent).

In butterflies and moths that feed little if at all after eclosion, the size of the abdomen relative to the size of the body has been used as a metric of reproductive effort, given that the abdomen contains all of the fat, reproductive organs, and nutrient reserves necessary to produce eggs. Quantified in this manner, reproductive effort for seven species of

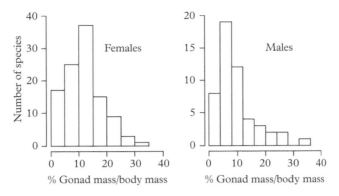

Figure 4.4 *One element of reproductive effort: proportional allocation of body tissue to gonadal tissue in fishes, expressed as percentages. Females: (mass of eggs at spawning) / (female mass) (n = 107 species). Males: (mass of testes at spawning) / (male mass) (n = 51 species). Data are from Wootton (1998).*

butterflies averages 46.4 per cent of body weight (se = 2.5 per cent) (Wickman and Karlsson 1989). Among other invertebrates, Roff (2002) reported a range in proportional allocations of 9–10 per cent for some species of starfish to as much as 75 per cent in the crab spider, *Misumena vatia*. For 22 species of British grasses, an average 30.9 per cent (se = 4.1 per cent) of the above-ground biomass is allocated to reproductive structures in the first flowering season (Wilson and Thompson 1989).

These examples make it clear that the proportional allocation of body mass to reproductive tissues can differ considerably. However, one needs to be mindful that these estimates are measured for individuals at the time they are breeding. In other words, they are proportional allocations of parental body mass measured per reproductive event. An interesting question is whether these differences among species remain when one accounts for differences in lifespan. More than 50 years ago, Williams (1966) predicted that reproductive effort should decrease as lifespan increases.

Williams' explicit accounting for lifespan was picked up by Charnov et al. (2007) who asked whether there was an empirical basis for asserting that lifetime reproductive effort might be roughly constant among species. Using the principles of metabolic scaling theory (which uses physiologically based processes associated with metabolic rate to explain patterns across broad taxonomic and geographic scales; Brown et al. 2004), they defined lifetime reproductive effort (LRE) as the mass of offspring that a female can produce during the course of her life, such that LRE = litter size × litters per year × adult lifespan × offspring mass at independence. Based on data for mammals and lizards, they found that, on average throughout their lives, females produce a mass of offspring approximately equal to 1.4 times their own body mass. As with the study of life-history invariants (sub-section 2.3.2), it is intriguing to think that overarching principles or processes, based perhaps on metabolic scaling principles, might be responsible for generating such consistencies among species.

4.4   Costs of Reproduction

The premise that reproduction requires energy allocations from a fixed pool of resources (Figure 4.3) leads to the logical conclusion that reproduction incurs costs to those elements from which energy is being diverted. If energy allocated to reproduction results in less energy being allocated towards survival, the probability of survival will decline. For indeterminately growing organisms, energy allocated to reproduction results in less energy being available for growth, potentially affecting future fecundity (Figure 4.2). Thus, an individual's decision to reproduce is predicted to exact costs to that individual's future probability of survival, reproductive success, and/or rate of growth (Williams 1966; Bell 1980).

There is widespread empirical evidence of costs of reproduction in plants and animals (Roff 1992, 2002; Stearns 1992). Taxonomically extensive reviews are available for reptiles (Shine 1980), plants (Obeso 2002), terrestrial mammals (Hamel et al. 2010), birds (Bleu et al. 2016), and fishes (Smith and Wootton 1995).

Costs have been detected through a variety of means. One approach is to plot a metric of future survival or reproductive success against a metric of current reproductive success; a declining relationship of some form is indicative of a cost (Figure 4.5(a)). An elegant study by Law (1979) involved a common-garden experiment on annual meadow grass (*Poa annua*). Plants in the experiment were grown from seeds removed from grasses inhabiting low-density sites ('opportunistic', first-colonizer plants) and high-density

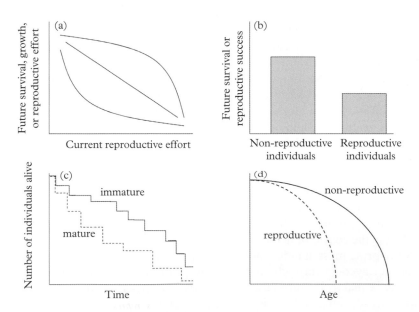

Figure 4.5 *Experimental outcomes or field observations consistent with the hypothesis that reproduction results in a cost to future survival or reproductive success.*

sites ('pasture' plants). Low-density plants, presumed to expend less energy on competition with other plants, were hypothesized to exhibit higher reproductive effort—and have shorter life expectancy—than high-density plants. Pooling data from opportunistic and pasture families, the number of inflorescences (a cluster of flowers, i.e. the seed-bearing part of a plant) produced in the second reproductive season was negatively related to the number of inflorescences in the first season, a pattern consistent with the existence of a cost (Figure 4.6(a)).

Another example from the plant literature is the finding that reproduction can reduce a plant's efficiency in using nitrogen for photosynthesis (the photosynthetic capacity of leaves is related to nitrogen content). A study of *Rhododendron lapponicum* found that one-year-old branches that had previously produced seed had 50–60 per cent of the efficiency of that of the same leaf generation of other branch types (Karlsson 1994).

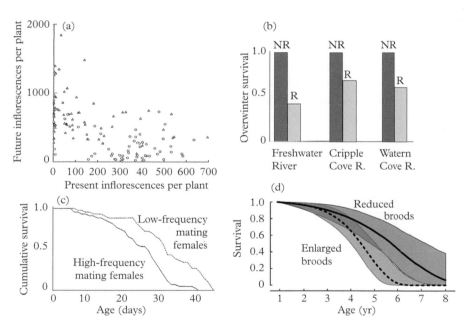

Figure 4.6 *Examples of research consistent with the hypothesis that reproduction bears costs to future survival and/or reproductive success: (a) future reproductive effort declines with increasing present effort in* Poa annua *(Law 1979) (triangles are pasture-family means, circles are opportunistic-family means); (b) reproductive (R) female brook trout have lower post-spawning, overwinter survival than non-reproductive (NR) females in three rivers (Hutchings 1994); (c) high mating frequency impairs longevity for female* Drosophila melanogaster *(Fowler and Partridge 1989); and (d) jackdaws with experimentally enlarged broods (steep, dashed line) experience a faster rate of decline in survival with age compared to those with reduced broods (shallow, solid line); shaded areas represent 95% confidence intervals (Boonekamp et al. 2014). Figure 4.6(a) is reprinted by permission from University of Chicago Press. Figure 4.6(d) is reprinted by permission from John Wiley and Sons.*

A conceptually simple approach to detecting costs is to compare survival probabilities between reproductive and non-reproductive individuals (ideally controlling for as many variables as possible that might also affect survival, such as age, size, and number of previous reproductive episodes) (Figure 4.5(b)). This approach can be undertaken in the field or in the lab. One such field study in Newfoundland, Canada, revealed evidence of survival costs in three populations of brook trout (*Salvelinus fontinalis*) (Figure 4.6(b)) (Hutchings 1994). The measurement of time in this type of work is binary, insofar as measurements are compared between two time periods ('before' and 'after'). An alternative approach, highly amenable to laboratory-based research, is to compare survival trajectories over continuous time (Figure 4.5(c)), a classic example of which compared survival 'curves' between female *Drosophila melanogaster* that mated at low and high frequencies (Figure 4.6(c)) (Fowler and Partridge 1989).

A fourth approach to studying costs involves manipulation of one element of reproductive effort, such as the addition or extraction of eggs in a bird's nest, followed by the monitoring of a metric of future reproductive success (survival, body condition) thereafter. Such manipulations can potentially reveal patterns of covariation between current effort and future survival, such as those depicted in Figure 4.5(d). However, many brood-manipulation studies encompass only a single breeding period, limiting the future time horizon across which costs might be realized and, thus, detected. A rare set of brood manipulations conducted throughout the life of a freely living vertebrate (jackdaws, *Corvus monedula*) found that mortality increased with age at a three-fold higher rate for birds with enlarged broods when compared with those with reduced broods (Figure 4.6(d)) (Boonekamp et al. 2014).

4.5   Energetic, Ecological, and Genetic Basis for Reproductive Costs

4.5.1   What constrains reproduction?

Proximate constraints associated with reproduction have potential to generate costs manifested in the short (acute costs) or long term (chronic or cumulative costs) (Table 4.1). Energetic and metabolic demands of reproduction constrain or limit resources that might otherwise have been available for factors related to survival, such as body maintenance and immunity. Ecologically generated constraints can also reduce the probability of future survival. Mating, for example, can elevate risk of predation; parental care can reduce short-term foraging opportunities, negatively affecting future growth and body maintenance. In addition to energetic and ecological constraints, costs may be generated by negative genetic correlations between traits affecting present reproduction and future survival. They might also be a product of genetic architecture, such that a gene is favoured by selection because of its positive effects on fitness early in life, despite having deleterious consequences to fitness later in life.

Table 4.1 *Hypothesized constraints and resultant costs of reproduction. An acute cost is realized in the immediate, short term; a chronic cost is paid over a longer time frame.*

Energetic constraints	Potential costs (A = acute, C = chronic)
Reduced ability to maintain basal metabolic rate because of allocations of resources to factors such as gonad production, mate competition, and parental care	Reduced survival during or immediately following breeding (A); reduced future fecundity or fertility (C); reduced future growth (C)
Increased risk of infection or parasitism due to weakened immune system	Reduced survival during or immediately following breeding (A) or later in life (C); reduced future fecundity or fertility (C); reduced future growth (C)
Increased risk of predation because of factors such as reduced locomotion, impaired vigilance, increased feeding rate	Reduced survival during or immediately following breeding (A)
Ecological and behavioural constraints	**Potential costs**
Increased vulnerability to predators and/or reduced feeding because of factors such as mate searching, mate attraction, and parental care	Reduced survival prior to, during, or immediately following breeding (A)
Physical injuries incurred during mate competition	Reduced survival prior to breeding (A) or later in life (C)
Genetic constraints	**Potential costs**
Antagonistic pleiotropy, i.e. a negative genetic correlation between traits associated with present and future survival and/or fecundity	Reduced survival and/or fecundity in the longer term (C)

4.5.2   Energetic constraints

When seeking a causal basis for costs, a logical first consideration is the energy demanded by reproduction. Energetic constraints can be attributed to the metabolic appropriation of carbohydrates, lipids, and proteins associated with various physiological (e.g. gonad production), morphological (e.g. seed protection), and behavioural (e.g. mating) correlates of reproduction.

Energetic and biochemical demands can be considerable. Egg mass alone comprises an average 12–13 per cent of the body tissue of fishes (Figure 4.4). For reptiles it is higher (19 per cent), notably so for snakes (26–30 per cent; Seigel and Fitch 1984; Seigel et al. 1986). Wenk and Falster (2015) concluded that seed production represented as little as 2 per cent of the total surplus energy (energy surplus to an organism's survival and maintenance requirements) in flowering plants of the southern-hemispheric Proteaceae (e.g. *Banksia* spp.) to as much as 53 per cent in subtropical woody dicots (e.g. *Acer* spp.). As discussed in section 4.3, these direct allocations to the production of gametes can be secondary to other demands of reproduction. When producing seeds, plants may expend resources necessary to mature a seed that do not involve the direct expense of provisioning the seed itself. From an energetics perspective, these accessory costs (per seed) have been estimated to account for 33 per cent to as much as 96 per cent of the total reproductive costs in some angiosperms (Lord and Westoby 2006).

In a study of butterflies, Wedell and Karlsson (2003) revealed not only substantive energetic expenditures associated with reproduction, but considerably different allocations between sexes. In the speckled wood (*Pararge aegeria*), females invest 71 per cent of body resources to reproduction as opposed to only 32 per cent for males. By contrast, energetic allocations are similar (57–58 per cent) for male and female green-veined white butterflies (*Pieris napi*). The sexual equivalence of energetic costs in this butterfly has been attributed to the observation that males, through the transmission of their spermatophores (protein capsules containing spermatozoa), provide nutrients to females during mating; speckled wood males do not (Wedell and Karlsson 2003).

Energetic constraints have consequences for future reproductive success, the most empirically demonstrable being reduced growth and, for many species, subsequently reduced fecundity (Figure 4.2). Good examples of reductions in future size associated with increased present reproductive effort have been documented in plants (Figure 4.7) and indeterminately growing vertebrates (Figure 4.8).

The energy demanded of reproduction has also been shown to indirectly affect future survival. One intriguing means is by increasing the risk of infection or parasitism due to a weakened immune system, a causal link that has been documented in birds, mammals,

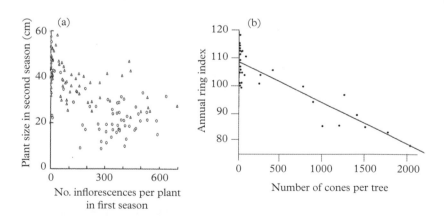

Figure 4.7 *Increased allocation of effort to reproductive structures reduces future growth in (a) annual meadow grass* (Poa annua) *(Law 1979) (triangles are pasture-family means, circles are opportunistic-family means) and (b) Douglas fir* (Pseudotsuga menziesii) *(Eis et al. 1965). In the latter example, the narrower the annual ring index, the slower the growth of the tree. Figure 4.7(a) reprinted by permission from University of Chicago Press. Figure 4.7(b) reprinted by permission from Canadian Science Publishing.*

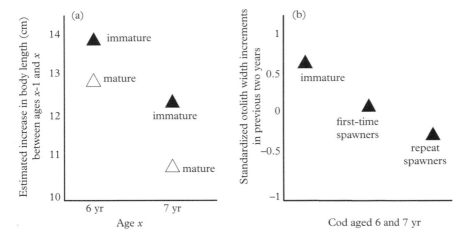

Figure 4.8 *Reproduction is associated with a slowing of growth in female Atlantic cod. (a) incremental increase in length during the previous year was less for mature cod, a finding supported by (b) incremental increases in the width of annuli in the otolith (bone used for ageing in fishes).*

Re-drawn from Folkvold et al. (2014).

lizards, and fishes (Pelletier et al. 2005). To take one example, a study of Soay sheep (*Ovis aries*) on Scotland's St. Kilda archipelago linked the increased parasite load caused by reproduction to lower future survival (Leivesley et al. 2019). In spring, females that give birth, and particularly those that wean a lamb, have significantly higher faecal egg counts of a parasitic, gastrointestinal nematode compared to females that do not reproduce. This increased level of parasitism is associated with lower body weight in summer and reduced survival during the subsequent winter.

Energetic constraints associated with reproduction can increase risk of predation by hindering an organism's escape response. The common whelk (*Buccinum undatum*), as do many molluscs, relies on a muscular 'foot' for locomotion. Brokordt et al. (2003) found that reproduction reduces metabolic capacity of the foot muscle in whelks, increasing their susceptibility to predation.

4.5.3 Ecological and behavioural constraints

Over and above the energetic allocations associated with reproduction, survival and fecundity costs can result entirely from factors external to an organism, generated by ecological and behavioural interactions with members of the same or other species.

Parental care, for example, by limiting mobility, can constrain parental behaviour to such an extent that it negatively affects survival or growth. This constraint has been well-documented in fishes for which costs are realized by the necessity of remaining in a restricted, poorly concealed spatial location that can increase predation risk and interrupt or severely restrict feeding (Smith and Wootton 1995). In reptiles, mobility can be limited by increased body mass. Gravid scincid lizards are considerably heavier than

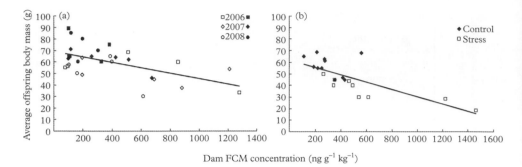

Figure 4.9 *Physiological stress caused by threat of predation can generate reproductive costs. Elevated, predator-induced levels of faecal cortisol metabolite (FCM) in female snowshoe hares is associated with reductions in offspring body mass. Source: Sheriff et al. (2009).*

Reprinted by permission from John Wiley and Sons.

non-gravid females and males, reducing locomotory speeds by 20–30 per cent and increasing their risk of being consumed by snakes (Shine 1980).

As noted above in Fowler and Partridge's (1989) study of *Drosophila* (Figure 4.6(c)), mating can prove costly to future breeding opportunities in arthropods. As with mammals and fishes, ecological constraints can be manifest by increased risk of parasitism and reduced foraging success; it can also cause genital damage (Arnqvist and Rowe 2005). In water striders (*Gerris buenoi*), simply the increased movements associated with mating can be sufficient to increase vulnerability to predation (Rowe 1994). Bright colouration, a secondary sexual characteristic in male guppies (*Poecilia reticulata*), can do the same (Houde and Endler 1990). A striking yet subtle means by which inter-specific interactions can affect individual costs has been documented in snowshoe hares (*Lepus americanus*). Predators, unsurprisingly, induce stress in potential prey. Sheriff et al. (2009) found that predator-induced increases in the concentration of glucocorticoids (steroid hormones related to physiological stress) can negatively affect both the size of a female's litter and the mass of her offspring (Figure 4.9).

Competition for mates, particularly when it involves behaviourally intense agonistic interactions, also has potential to increase mortality. Although such conflicts are often settled without contact (through a range of displays and cues that transfer information between individuals, communicating their relative chances of success), physically aggressive encounters do occur. Outcomes can be severely problematic for losers when the competitive abilities of the interacting individuals are highly asymmetric. Physical interactions between large (>50 cm) and small (<15 cm) male Atlantic salmon competing for access to females during spawning can result in life-threatening and life-ending injuries to the smaller males (Hutchings and Myers 1987; Fleming 1996).

4.5.4   Senescence

Based on the empirical literature, it would be fair to conclude that most reproductive costs are realized in the short term (relative to the lifespan of an individual), being manifest before, during, or shortly after breeding. However, some costs might not be expressed until long after maturity, when an individual has reached older ages. Such chronic or cumulative costs can represent important contributors to actuarial senescence (the rate at which mortality increases with age) and/or reproductive senescence (the rate at which female fecundity or male fertility declines with age).

As noted in Figure 4.6(d), Boonekamp et al. (2014) found that experimentally enlarged brood sizes were associated with an increase in the *rate* at which mortality increased with age in jackdaws. Across many species of mammals, the onset of actuarial senescence has been attributed to physiological mechanisms associated with sexual selection, potentially accounting for differences in longevity between males and females (Tidière et al. 2015). Evidence for reproductive senescence has also been reported in some fishes (e.g. Atlantic herring; Benoît et al. 2018) and in at least one insect (European field cricket, *Gryllus campestris*; Rodríguez-Muñoz et al. 2019).

In addition to negative genetic correlations, reproductive costs that lead to actuarial and reproductive senescence have been hypothesized to originate from the actions of a single gene acting on more than one fitness-related trait (sub-section 3.2.3). The idea is that genes that have a positive influence on a trait in early adult life (such as fecundity) might be favoured by natural selection even though the same genes have a negative effect on a fitness-related trait (such as survival) later in adult life. This 'antagonistic pleiotropy' was first proposed by Williams (1957) as a possible explanation for the evolution of ageing or senescence. Laboratory selection experiments have since provided evidence of antagonistic pleiotropy. A classic one was undertaken by Rose (1984) who found that selection for early fecundity in *Drosophila* was associated with short lifespans, while selection for late fecundity was correlated with longer lifespans.

4.5.5 Measuring effort and costs: challenges and caveats

A fundamental assumption of life-history theory is that greater reproductive effort exacts greater reproductive costs. Logically, this makes sense. However, empirically demonstrating this relation has sometimes proven problematic. It need not be so. One simply needs to be reminded that a life-history trade-off is a *within-individual* phenomenon. Think back to the allocation trade-offs depicted in Figure 4.3. Differential allocations to growth, survival, and reproduction are made *within* individuals. Antagonistic pleiotropy is a product of gene interactions that occur *within* an individual's genome.

This is not to say that one cannot draw conclusions about the existence, or the magnitude, of costs by comparing metrics of reproductive effort and their consequences among different individuals. You can indeed get a *sense*, or an approximation, of the magnitude of costs by comparing reproductive with non-reproductive individuals, or individuals with artificially enlarged versus reduced clutch sizes. But when doing so you must control for as many confounding variables as possible. Whatever the resulting correlations in an among-individual comparison, bear in mind that the analysis is a conservative one insofar as it is likely to underestimate, rather than overestimate, costs.

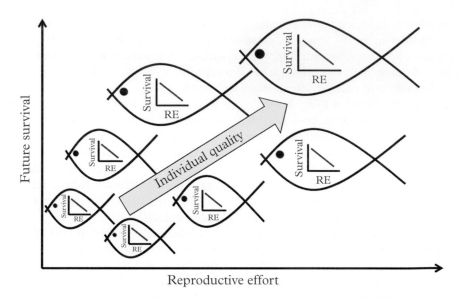

Figure 4.10 *Trade-offs between present reproductive effort (RE) and future survival are negative associations that exist within individuals, represented by red lines between survival and RE within each diagrammatic fish. Many studies of reproductive costs compare different individuals. Some of these have uninformatively yielded positive associations between future survival and effort, such as that indicated by the 'fish' here. This is often because within-individual costs of reproduction have been masked by individual differences in quality.*

When comparing different individuals, a classic problem in relating costs to effort arises when the individuals under study differ in some metric of quality. Individuals might (and usually do!) differ from one another because of things such as physiological condition, learning ability, metabolic rate, foraging efficiency, predator avoidance, etc. Thus, under some circumstances, a plot between current reproductive effort and a measure of future reproductive success can mask a true negative relationship or yield a positive association, even though reproductive costs exist within each individual (Figure 4.10).

The obvious caveat to such among-individual comparisons is that one cannot defensibly conclude from such studies that a survival or fecundity cost does not exist. To accurately quantify the reproductive costs experienced by a given individual, one needs to be able to compare the future conditions (at age $x + t$) experienced by that individual (in terms of survival, fecundity, or growth) depending on whether it had reproduced in the past (at age x) or not.

A second challenge in evaluating the relationship between effort and cost is the likelihood that the magnitude of a cost and the probability of its detection almost certainly depend on the stressfulness of the environmental conditions under which the organisms are studied. Costs are more likely to be detected in stressful rather than benign environments, a prediction supported, for example, by work on *Drosophila melanogaster* (Marshall and Sinclair 2010). In sum, the estimation of reproductive costs can be very challenging, despite the logical basis for their existence.

4.6   Summing Up and a Look Ahead

Trade-offs are at the core of life-history theory. Predictions about life-history evolution are intellectually bereft without their consideration. Reproductive effort is the proportion of total energy or resources allocated to all elements of reproduction. The elements of effort are wide-ranging: energy to produce seeds or eggs; pollinator-attraction mechanisms; fruit production; flowers; migration; mating; parental care. Reproductive effort generates reproductive costs. Increases in current reproductive effort reduce future reproductive success by affecting survival, growth, and/or fecundity. The causal mechanisms of these costs can be energetic, ecological, behavioural, or genetic. Evidence for reproductive costs is widespread. Instances where the evidence of costs is equivocal are usually caused by using among-individual correlations to study what is a within-individual phenomenon.

This chapter is the fourth to focus on a foundational element of life-history theory. These elements (population dynamics, phenotypic variation, genetic underpinnings, reproductive effort and costs) are necessary if one hopes to profitably understand and engagingly explore the evolutionary grandeur of life-history variability. Chapter 5 provides a transition, offering a method for exploring how the joint contributions of survival and fecundity to fitness can be estimated in quantitatively accessible ways.

5

Vital Rates

5.1 Schedules of Mortality and Reproduction

Surviving to maturity and reproducing thereafter are obvious prerequisites for a suc-
cessful life history. However, given the enormous potential variability in life-history traits
(Chapter 2), some realized life histories will inevitably be more successful than others,
depending on a phenotype's genetic background and the environment in which it lives
(Chapter 3). But what constitutes success? Here, we return to the concept introduced in
Chapter 1 of fitness, the primary currency of ecology, evolution, and adaptation.

As a reminder, the success of the life history of an individual is determined by the
fitness associated with that life history, relative to the fitness of alternative life histories
expressed by other members of the same population. As discussed in Chapter 1 and,
more specifically in sub-sections 2.4.5 and 2.4.6, fitness is often defined as r or, more
precisely, r_{max}. Fitness is a function of how survival and reproductive effort change
through life. Williams (1966) identified the optimal schedule of allocation to reproductive
effort throughout life as a key overarching determinant of an organism's life history.

These age-determinant schedules are defined by age-specific rates of survival (l_x) and
fecundity (b_x). Age-specific survival, l_x, is the probability of surviving (expressed as a
decimal proportion) from birth until the beginning of age x. b_x is the number of offspring produced by an individual breeding at age x. Across taxa, three general patterns
of survival with age have been identified; the chapter begins with descriptions of these
survival types. Fecundity tends either to remain constant or to increase with age until the
organism begins to senesce, depending on whether growth is determinate or indeterminate. Life tables, which provide a logistically tractable means of expressing l_x and b_x, are
then introduced.

The chapter then compares three classic definitions of fitness in a life-history context:
the net reproductive rate (R_0); the intrinsic rate of increase (r); and reproductive value
(RV). Having established that l_x and b_x are fundamental to estimating each of these
measures, the chapter explores how life tables can be used to estimate fitness.

Finally, the chapter picks up on Williams' (1966) idea of an optimal schedule of allocation to reproductive effort throughout life. Here, life table analyses are used to explore
how the costs of reproduction discussed in Chapter 4 can influence the optimal age at
maturity, i.e. the age at maturity that maximizes fitness.

A Primer of Life Histories: Ecology, Evolution, and Application. Jeffrey A. Hutchings, Oxford University Press. © Jeffrey A. Hutchings 2021.
DOI: 10.1093/oso/9780198839873.003.0005

5.2   Life Tables

5.2.1   Age-specific survival and fecundity

In a life-history context, fitness is a function of lifetime schedules of reproductive effort reflected by age-specific survival (l_x) and fecundity (b_x). Generically, these are often termed 'vital rates'. Although x usually denotes age, it can also be used to represent a life stage—such as egg, pupa, larva, adult—rather than age. For simplicity, and because it is the more commonly applied unit of measurement of x, in this book l_x and b_x will refer to age-specific survival and age-specific fecundity, respectively, unless otherwise indicated.

The study of age-specific survival has a long history because of its unambiguous relationship with the probability that a human will live from one age to the next. Estimates of expected lifespan from birth, or from later ages in life, are essential for actuarial companies to guide them in establishing life-insurance policies and rates. The current practice of using l_x to denote age-specific survival originated with the work of Gompertz (1820, 1825, 1861), a mathematician and early actuarial researcher. In his 1820 and 1825 papers on 'life contingencies', he used L_x to express 'the number of persons who would be living at the age x, out of the number of persons who may have been living at some given common previous age' (Gompertz 1861: 390). This is essentially the same definition as l_x, expanded to include all organisms, not only humans.

As documented by Gompertz and many others, humans (and other large mammals) express a pattern of age-specific survival characterized, on average, by relatively high survival in early and middle life but rapidly declining l_x later in life. This relationship between l_x (accentuated on a log scale) is called a Type I survivorship curve. Originally defined by Pearl and Miner (1935) and refined by Deevey (1947), survivorship curves illustrate general patterns of change in l_x with age (Figure 5.1). The Type II survivorship

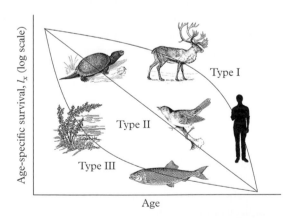

Figure 5.1 *Three survivorship curves showing relationships between age-specific survival probabilities from birth (l_x, log scale) as a function of age. The Type I curve applies to organisms such as large mammals, including humans. The Type II function applies to many reptiles and birds. The Type III curve is characteristic of species that experience very high mortality early in life, such as many plants and fishes.*

curve is characteristic of species that have a constant rate of mortality throughout their lives (e.g. some turtles and birds). Many species exhibit a Type III survivorship curve, characteristic of species for which mortality is exceedingly high in early life until individuals reach an age when their risk of death declines substantially, after which they experience relatively high survival. Examples include plants with wind-dispersed seeds and broadcast-spawning marine invertebrates and fishes.

Patterns of age-specific fecundity tend to follow one of two forms. For indeterminately growing organisms that continually increase with size as they age (e.g. plants, fishes, amphibians, reptiles; sub-section 2.2.8), the number of offspring produced per individual per unit of time increases with age because of the positive association between individual size and numbers of seeds or eggs (e.g. Figure 4.2(b)). Among determinately growing organisms, such as mammals, birds, and many insects (e.g. flies, beetles, moths), fecundity tends to remain roughly constant with increasing age. For organisms that live long enough, age-specific fecundity declines as organisms begin to senesce.

5.2.2 An example of a life table

For illustrative purposes, consider a life table of vital rates for the black-capped chickadee (*Poecile atricapillus*) (Table 5.1). The number of individuals alive at each age x (n_x) is used to calculate survival from one age to the next (s_x) and survival from birth to age x (l_x), such that $s_x = n_x/n_{x-1}$ and $l_x = n_x/n_0$.

Depending on the organism, the survival data might be based on direct counts of individuals (comparatively straightforward for plants, large mammals, and birds). Alternatively, when direct counts are difficult to obtain, estimates can be made from

Table 5.1 *Life table data for the black-capped chickadee, a North American member of the family Paridae (small passerine birds that are widespread in Europe, North America, Africa, and Asia). Values of l_x are based on Loery et al. (1987), Albano (1992), Smith (1995), and Ramsay et al. (2003). Values of b_x are based on Mahoney et al. (1997).*

Age (x)	Number alive (n_x)	Survival from age $x-1$ to x (s_x)	Age-specific survival, l_x ($= l_{x-1} \times s_x$)	Age-specific fecundity, b_x
0	1000	1.000	1.000	0
1	190	0.190	0.190	6
2	140	0.737	0.140	6
3	90	0.643	0.090	6
4	60	0.667	0.060	6
5	40	0.667	0.040	6
6	20	0.500	0.020	6
7	0	0	0	—

mark-recapture studies. This involves marking a group of individuals of a given age x and re-sampling the same population at one or more later time intervals to determine the proportion that survived. Fecundity data (b_x) are based on average values for individuals at a specific age.

If the l_x values are roughly constant through time, the relative number of individuals in each age class will be similar from one generation to the next. This results in a 'stable' age distribution. For example, the ratios of the numbers of individuals aged 1 through 6 (i.e. the age distribution), beginning with 190 ÷ 20, in Table 5.1 are 9.5 : 7.0 : 4.5 : 3.0 : 2.0 : 1.0. If this age distribution is stable through time, these ratios will not change. Populations with a stable age distribution need not be numerically stable (i.e. stationary) through time; the populations might also be increasing or decreasing. Life tables provide one of the most easily understood means of estimating fitness (section 5.4). In terms of age structure, the longer the period of time that an increasing, decreasing, or stationary population expresses a stable age distribution, the greater the temporal stability of that estimate of fitness.

5.3   Fitness

5.3.1   Rates of increase

The intrinsic rate of increase (r) and its analog the Malthusian parameter (m) have been explicitly linked with the concept of fitness since the early twentieth century. The Malthusian parameter was central to Fisher's thinking in the 1930s, reflected by his unequivocal assertion that 'm measures fitness' (Fisher 1930: 34). He recognized that m, and the vital 'statistics' (l_x and b_x) that determine m, measure both a population's per capita rate of increase and individual fitness:

> The vital statistics of an organism in relation to its environment provide a means of determining a measure of the relative growth-rate of the population, which may be termed the Malthusian parameter of population increase, and provide also a measure of the reproductive values of all individuals at all ages or stages of their life-history. The Malthusian parameter will in general be different for each different genotype, and will measure the fitness to survive of each. (Fisher 1930: 46)

Thus, fitness is a property of genotypes, or individuals, not populations and not species. The higher the fitness of an individual, the greater its ability to propagate its genes to future generations. Fisher's text might seem to confuse or conflate the rates of increase of populations and individuals. This can be reconciled by thinking of the fitness of an individual i as r_i, and the population's per capita rate of growth as the average value of r_i amongst all of the individuals in that population.

For a variety of reasons, the use of m declined through time in favour of r. Mathematically, this was not problematic because Fisher's (1930) definition of m was the same as Lotka's (1907) definition of r; both are equal to $b - d$ (Equation 1.8).

The concept of fitness as a *rate of increase* is very useful and can broaden the utility of the parameter r. Take, for example, a mutation that affects a particular gene or group of genes for which there are consequences for an individual's age-specific survival or

fecundity. If the mutation influences an individual's birth or death rate, it will by necessity affect r (except for the unlikely scenario in which the proportionate changes to birth
and death rates are equal). So, r can be used as a means of studying the rate at which a
gene substitution (insofar as it benefits an individual's fitness) spreads through a population (Charlesworth 1980). If, as suggested earlier, the value of r for a given population
reflects the average fitness of individuals within that population, then changes in per
capita population growth rate can be said to reflect changes in the average fitness of
individuals within that population.

5.3.2   Net reproductive rate, R_0

Before turning to how r can be measured, consider first a simpler measure of fitness, the
net reproductive rate or R_0. Often described as 'lifetime reproductive success', this metric represents the average number of offspring produced over the lifetime of an average
individual. It is readily calculated from a life table. For a particular life history, at each
age x, multiply l_x by b_x and sum the $l_x b_x$ values, such that R_0 is:

$$R_0 = \sum_0^\infty l_x b_x$$

Equation 5.1

Because R_0 is calculated over an entire lifetime, it is a per-generation parameter (unlike
r which measures the rate of increase per unit of absolute time, irrespective of generation
length).

For sexually reproducing organisms, R_0 is often defined as the average number of female
offspring produced during the lifetime of a female (in many species, it is often females
that are more limited in the ability to reproduce). If the lifetime production of offspring
by a female (discounted in the life table by the probabilities of living to each age) is such
that she is exactly able to replace herself, then $R_0 = 1$. If the average R_0 among individuals in the population is equal to 1, the population is stationary. If it is less than or greater
than 1, the population is exponentially decreasing or increasing, respectively.

5.3.3   Intrinsic rate of increase, r

The intrinsic rate of increase, r, can be calculated for populations that are numerically
changing continuously or at discrete intervals through time. For a population growing or
declining exponentially and continuously,

$$1 = \int_0^\infty e^{-rx} l_x b_x dx$$

Equation 5.2

For a population growing or declining exponentially at discrete intervals,

$$1 = \sum_0^\infty e^{-rx} l_x b_x$$

Equation 5.3

In this book, Equation 5.3 will be used because it tends to be simpler to understand and
because it reasonably captures the life histories of many organisms for which births and/
or deaths occur at comparatively discrete intervals.

Equation 5.3 is the Euler-Lotka equation, named after Leonhard Euler, an eighteenth-century Swiss mathematician who was the first to use the exponential function and logarithms in analytical proofs, and Alfred Lotka, whose mathematical studies on populations were introduced in Chapter 1 with respect to r. The parameter r is constant across all ages, whereas l_x and b_x can, and normally do, change with age. The Euler-Lotka equation is used to solve for r. To do so, an iterative process is required. This means that the unknown parameter r is solved by a process of trial and error that involves inserting different values of r into the Euler-Lotka equation to determine which one renders the right-hand side of the closest to one.

It is quite likely that it will not be intuitive why the solution for the Euler-Lotka equation necessitates having the right side of the equation equal 1. Before going further, it might be helpful to derive the equation to help understand its final structure (i.e. Equation 5.3). Begin by defining the number of newborn individuals at any time t born to females of age x as:

$$n_{newborn(t)} = n_{newborn(t-x)} \times l_x \times b_x$$

Equation 5.4

where $n_{newborn(t)}$ is the number of newborn offspring at time t, $n_{newborn(t-x)}$ is the number of newborn x time units ago, l_x is the probability of surviving from birth to age x, and b_x is the number of newborns produced by a female at age x. For example, if all individuals reproduce only at age 3, the equation states that the number of newborn offspring this year (i.e. year t) is equal to the number of newborn offspring born three years ago multiplied by the probability of those offspring surviving until age 3 (i.e. l_3) and the average number of offspring produced by females at age 3 (i.e. b_3).

Of course, breeding groups are typically comprised of individuals of multiple ages. So, we need to account for the contributions of offspring produced by females of different ages and the probabilities that those females survived from their birth x years ago until year t. Summing across several ages, ranging from age at maturity (α) to age at death (ω), we have:

$$n_{newborn(t)} = \sum_{\alpha}^{\omega} n_{newborn(t-x)} \, l_x \, b_x$$

Equation 5.5

Recall that the model for exponential population growth (Equation 1.11) is $N_t = N_0 \, e^{rt}$. Replacing N_0 with $n_{newborn(t-x)}$, we can write:

$$n_{newborn(t)} = n_{newborn(t-x)} \, e^{rx}$$

or

$$n_{newborn(t-x)} \, e^{rx} = n_{newborn(t)}$$

Equation 5.6

Dividing both sides by e^{rx} yields:

$$n_{newborn(t-x)} = n_{newborn(t)} / e^{rx}$$

or

$$n_{newborn(t-x)} = n_{newborn(t)}\, e^{-rx}$$

Equation 5.7

Substituting Equation 5.7 into Equation 5.5 yields:

$$n_{newborn(t)} = \sum_{\alpha}^{\omega} n_{newborn(t)}\, e^{-rx}\, l_x\, b_x$$

Equation 5.8

Finally, dividing both sides by $n_{newborn(t)}$ yields the Euler-Lotka equation:

$$1 = \sum_{0}^{\infty} e^{-rx}\, l_x\, b_x$$

As this derivation indicates, the '1' on the left side of the equation carries no more significance than being a means to simplify an equation (in this case, Equation 5.8).

Returning to the model for continuous exponential population growth, $N_t = N_0\, e^{rt}$, the equation readily tells us the value of r at which a population is stable over time, i.e. when $N_t = N_0$. This occurs when $r = 0$. A positive value of r is indicative of a population that is increasing exponentially; a negative value of r is indicative of a population that is decreasing exponentially.

As noted in Chapter 1, changes in population size (N) from one discrete time step (t) to the next time step ($t + 1$) can be represented by the finite (or geometric) rate of population growth, λ, such that $N_{t+1} = \lambda N_t$ (Equation 1.4) and $\lambda = N_{t+1}/N_t$ (Equation 1.5). λ is a discrete measure of population growth whereas r is a continuous measure of per capita population growth. As a measure of fitness, λ can be thought of as the average contribution of each individual alive at time t to the size of the population at time $t + 1$. It differs from the intrinsic rate of increase insofar as r is the average contribution of each individual to the *rate of change* in population size. The two parameters are mathematically equivalent when the time step between t and $t + 1$ is infinitesimally small, such that $\lambda = e^r$ and $r = \ln(\lambda)$ (see sub-section 1.2.1). As with R_0, when $\lambda = 1$, the population is not changing in abundance through time (i.e. $N_{t+1} = N_t$). If a population is exponentially decreasing or increasing, it implies that λ is less than or greater than 1, respectively.

5.3.4   Reproductive value, *RV*

A third measure of fitness is reproductive value. Much less commonly used than r and R_0, reproductive value at age x represents the present and future production of offspring by an individual breeding at age x and living through its maximum possible life span, discounted by the probability of that individual surviving to its oldest potential age. Reproductive value, RV_x, in a stationary population (i.e. $R_0 = 1$ or $r = 0$) can be calculated as:

$$RV_x = \sum_{t=x}^{\infty} (l_t b_t)/l_x$$

Equation 5.9

Fisher (1930), who introduced the term, described *RV* as the extent to which individuals of a given age will, on average, contribute to the ancestry of future generations. Williams (1966) partitioned reproductive value into present and residual reproductive value, explicitly acknowledging that greater effort expended at age x will have consequences for the effort, and the reproductive value, that can be expended in the future.

The concept of reproductive value is instructive when thinking about how selection acts on the level of effort that an individual expends at age x, given its probability of surviving to and contributing offspring at future ages. However, the resource partitioning that maximizes the contribution to future generations will also depend on whether the population is stable, growing, or declining. In a growing population, for example, selection should favour higher levels of reproductive effort at younger ages because it would increase the rate at which one's genes are proportionally represented in the expanding population. Being a within-population measure of relative fitness, RV can be used to determine which age (or size) classes make the highest contributions to the total reproductive output of a cohort or year class.

5.3.5   Caveats

Which measure of fitness should one use? The answer can depend on whose papers you read and the clarity with which those papers have been written. Notwithstanding some opacity, one thing that is clear is that the metrics described here share the feature that each depends on l_x and b_x. As fitness measures, it is also clear that r, R_0, and RV differ in terms of the extent to which they can be used to compare rate of increase among members of the same population, different populations, or different species.

When considering fitness from a life-history perspective, think of it as a measure of an individual's lifetime reproductive success and the rate at which that success is translated by reproductively competent individuals to future generations. There can be advantages in selecting a measure of fitness that reflects both; in essence, a 'rate of change' of reproductive success. In this regard, r and R_0 have advantages over reproductive value. Furthermore, r and R_0 can be used to determine the direction (increase, decrease) and rate (fast, slow) of temporal changes in population size. Quite importantly, they can be used comparatively among populations.

Estimates of reproductive value (RV), on the other hand, while being meaningful to the population under study, cannot be directly compared among populations. However, it can be a useful parameter to estimate when evaluating the relative importance of individuals of different ages and sizes to population stability and growth (Hutchings and Rangeley 2011; Kindsvater et al. 2016).

While r and R_0 both provide information on directional change in population size, the time frames of the two parameters differ (always critical when thinking about rate of change). Recall that R_0 is a per-generation rate of change. If you are comparing the net reproductive rates of two different populations of the same species that also differ in generation time, you cannot use R_0 to compare changes in population size from one year to the next, only from one generation to the next. To compare growth rates of populations that differ in generation time, we need to estimate r. There is also the advantage that, as a rate of increase, r can be directly compared among different populations and even different species.

To some extent, the choice of fitness can depend on the study organism and on the relative ease with which data on survival and fecundity can be measured under natural conditions; it is not unusual for field biologists, for example, to favour R_0. It can also depend on whether the primary interest is in measuring fitness per se, or whether there is interest in translating average individual fitness to metrics of population viability, growth, or sustainable harvesting, in which case r is often favoured. Lastly, in a modelling context, the choice of r or R_0 can depend on how tractable the parameters are analytically and the ease with which they can account for variability in life-history traits, such as body size. Under these circumstances, R_0 is favoured by some researchers over r.

Given its widespread use and historical scientific legacy, this book will primarily use r to measure fitness, being mindful that there are instances in which the measure of fitness used can occasionally lead to different conclusions (Brommer 2000).

5.4   Estimating Fitness from a Life Table

As introduced by Fisher (1930), life tables offer a simple and straightforward means of estimating fitness within a life-history context. The net reproductive rate, for example, can be readily calculated, as illustrated in Table 5.2 by expanding the number of columns in the life table previously presented for chickadees (Table 5.1). Recall that:

$$R_0 = \sum_0^w l_x b_x$$

The $l_x b_x$ products are presented in column 6 of Table 5.2. Their sum is 3.24. Thus, for a population whose vital rates correspond to those in Table 5.2, each individual produces 3.24 individuals that will survive and reproduce in the next generation. The population will increase over time because each individual is producing, on average, more individuals than is necessary to replace itself ($R_0 > 1$). Generation time (G), defined as the average age of the parents of a single 'cohort' or 'year class' (e.g. all of the young born in 2001), can be calculated as $\sum l_x b_x \, x / \sum l_x b_x$. For the chickadee life history reflected by Table 5.2, generation time is $(7.80/3.24) = 2.41$ years (see columns 7 and 6).

As noted in sub-section 5.3.5, if you are interested in the per capita rate of population growth for a single population, R_0 provides an appropriate metric. The net reproductive rate can also be compared among populations, or even among species, but only if their generation times are the same, given that R_0 is a per-generation rate of increase whose units are 'individuals per individual per generation'. If you wish to compare per capita population growth among populations or species that differ in generation time, you need to estimate r, whose units are 'individuals per individual per instantaneous change in time'.

Table 5.2 *Life table data for the black-capped chickadee, expanded to include additional columns that clarify the calculations of R_0 and r. (Calculations rounded to three decimal places.)*

Column Numbers

1	2	3	4	5	6	7	8	9	10
Age (x)	Number alive (n_x)	Survival from x − 1 to x (s_x)	Age-specific survival (l_x)	Age-specific fecundity (b_x)	$l_x b_x$	$l_x b_x$ x	$l_x b_x$ e^{-rx} r = 0.49	$l_x b_x$ e^{-rx} r = 0.70	$l_x b_x$ e^{-rx} r = 0.61
0	1000	1.000	1.000	0	0	0	0	0	0
1	190	0.190	0.190	6	1.14	1.14	0.698	0.566	0.619
2	140	0.737	0.140	6	0.84	1.68	0.315	0.207	0.248
3	90	0.643	0.090	6	0.54	1.62	0.124	0.066	0.087
4	60	0.667	0.060	6	0.36	1.44	0.051	0.022	0.031
5	40	0.667	0.040	6	0.24	1.20	0.021	0.007	0.011
6	20	0.500	0.020	6	0.12	0.72	0.006	0.002	0.003
7	0	0	0	—					
					$\Sigma =$ 3.24	$\Sigma =$ 7.80	$\Sigma =$ 1.216	$\Sigma =$ 0.870	$\Sigma =$ 1.000

There are two ways of estimating *r* from a life table (note that the estimates of *r* from a life table represent the realized per capita rate of increase, $r_{realized}$; see sub-section 1.2.3 and Figure 1.4). The first provides an approximation, based on R_0, by correcting for differences in generation time. This is achieved by dividing the natural logarithm of R_0 by generation time, such that *r* can be approximated as $\ln(R_0)/G$, yielding $r \sim 0.49$ for the life-history data in Table 5.1.

A more accurate calculation of *r* is obtained from the Euler-Lotka equation (Equation 5.3):

$$1 = \sum_{0}^{\infty} e^{-rx} l_x b_x$$

Age and age-specific survival (l_x) and fecundity (b_x) are known from the life table. The unknown parameter *r* can only be solved by iteration, that is, by inserting different values of *r* into the Euler-Lotka equation to determine which one renders the right-hand side of the equation closest to 1. (As an aside, this is an example of what was meant by differences in mathematical tractability at the end of sub-section 5.3.5. Use of the Euler-Lotka equation to estimate *r* requires iteration, whereas calculation of R_0 does not require iteration.)

For the chickadees, we can begin by inserting the approximation of r (0.49) obtained by dividing $\ln(R_0)$ by G. The right-hand side of the Euler-Lotka equation is equal to 1.216 (column 8, Table 5.2). This is close to 1 but not as close as it could be. Increasing the value of r to 0.70 yields $\sum l_x \, b_x \, e^{-rx} = 0.870$ (column 9, Table 5.2). Through the trial-and-error process of iteration, we eventually find that $\sum l_x \, b_x \, e^{-rx}$ equals 1 when $r = 0.61$ (column 10, Table 5.2) (rounding calculated numbers to three decimal places).

5.5 Life-Table Approach to Exploring Optimal Age at Maturity

5.5.1 Incorporating a fecundity cost of reproduction

Life tables offer a straight-forward and accessible means of exploring how trade-offs can influence life histories. Integral to such exploratory analyses is the concept of optimality. The optimal life history is that which generates the highest per capita rate of growth, i.e. the highest values of r, or fitness, relative to potentially alternative life histories in the same population.

The data in Table 5.3 represent age-specific rates of survival and fecundity for an indeterminately growing organism for which there are four potential life histories. These four life histories are represented by four ages at maturity (α), i.e. $\alpha = 2, 3, 4$, and 5 years and four age-specific schedules of fecundity, b_x. Age-specific survival schedules are the same for each α, but age-specific fecundity changes with α. These changes in fecundity reflect a fecundity cost of reproduction; individuals maturing at age α allocate less energy to future growth because of their energy allocation to reproduction (see Figures 4.2 and 5.2). Thus, those that mature at age 2 have a lower fecundity at age 5 ($b_5 = 800$) than individuals that mature at age 5 ($b_5 = 4000$) (Table 5.3).

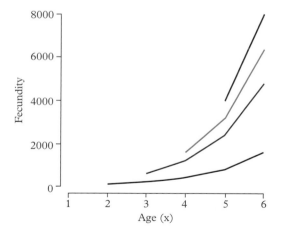

Figure 5.2 *Fecundity as a function of age for individuals maturing at age 2 (black), 3 (red), 4 (green), and 5 (blue), in accordance with life tables in Tables 5.3 and 5.4.*

Given that there are four possible ages at maturity in this population, the question arises as to which age is the optimal age at maturity (α_{opt}), i.e. the α associated with the highest fitness and, thus, the α that selection would be expected to favour. Using the Euler-Lotka equation to estimate r by iteration (as done in section 5.4), we find that α_{opt} is 3 yr (Table 5.3).

Table 5.3 *Simulated life table data for an indeterminately growing organism for which fecundity increases with age. Age at maturity is α. The fitness associated with each age at maturity is indicated by r_α. The life table incorporates a fecundity cost of reproduction; an individual maturing at age 2 has a lower value of b_5 and b_6 than individuals maturing at ages 3, 4, or 5.*

Age (x)	Survival from $x - 1$ to x (s_x)	Age-specific survival, l_x ($l_{x-1} \times s_x$)	Age-specific fecundity, b_x
0	1	1	0
1	0.2	0.2	0
$\alpha = 2$	0.1	0.02	100
3	0.5	0.01	200
4	0.5	0.005	400
5	0.5	0.0025	800
6	0.5	0.00125	1600
$r_{\alpha=2} = 0.682$			
0	1	1	0
1	0.2	0.2	0
2	0.1	0.02	0
$\alpha = 3$	0.5	0.01	600
4	0.5	0.005	1200
5	0.5	0.0025	2400
6	0.5	0.00125	4800
$r_{\alpha=3} = 0.785$			
0	1	1	0
1	0.2	0.2	0
2	0.1	0.02	0
3	0.5	0.01	0

Age (x)	Survival from $x-1$ to x (s_x)	Age-specific survival, l_x $(l_{x-1} \times s_x)$	Age-specific fecundity, b_x
α = 4	0.5	0.005	1600
5	0.5	0.0025	3200
6	0.5	0.00125	6400
$r_{\alpha=4} = 0.664$			
0	1	1	0
1	0.2	0.2	0
2	0.1	0.02	0
3	0.5	0.01	0
4	0.5	0.005	0
α = 5	0.5	0.0025	4000
6	0.5	0.00125	8000
$r_{\alpha=5} = 0.552$			

5.5.2 Incorporating a survival cost of reproduction

The life table in Table 5.3 provided an opportunity to calculate the optimal age at maturity albeit under a fairly restrictive set of alternatives. Age-specific fecundity could take different values, depending on age at maturity (α), to reflect a fecundity cost of reproduction. But the fecundity cost itself (i.e. the extent to which future fecundity was diminished by age at maturity) did not vary.

To make the life table more flexible, we can expand Table 5.3 to include two additional columns (Table 5.4). The second column in Table 5.4 allows for inclusion of a survival cost of reproduction (*survcost*), ranging from zero (no cost) to one (death). The survival cost begins to be expressed during the year immediately following the age at maturity. Thus, if α = 2, the cost is expressed at age 3 and every year thereafter that reproduction takes place. The survival cost can be incorporated into the calculations of l_x by simple multiplication. Table 5.4 illustrates the mechanics of this calculation.

The expansion in Table 5.4 also includes an option to account for new sources of extrinsic mortality (*newmort*). (The 'new mortality' is additional to the mortality that the population already experiences, as reflected by age-specific declines in l_x.) This option can be useful when modelling how optimal age at maturity might change under increasing mortality at one or more ages because of natural or human-induced environmental change. This will be explored further in sub-section 5.5.3.

Table 5.4 *Simulated life table data for an indeterminately growing organism given in Table 5.3, expanded to accommodate a survival cost of reproduction (*survcost; *column 2) and proportional increases to natural mortality (*newmort; *column 3). The terms* survcost *and* newmort *range between zero and one. In this table, both* survcost *and* newmort *are set to zero.*

Age (x)	(1 − survcost) (A)	(1 − newmort) (B)	Survival from x − 1 to x (s_x) (C)	Age-specific survival (l_x) (l_{x-1} × A × B × C)	Age-specific fecundity (b_x)
0	1	1	1	1	0
1	1	1	0.2	0.2	0
α = 2	1	1	0.1	0.02	100
3	1	1	0.5	0.01	200
4	1	1	0.5	0.005	400
5	1	1	0.5	0.0025	800
6	1	1	0.5	0.00125	1600
$r_{\alpha=2} = 0.682$					
0	1	1	1	1	0
1	1	1	0.2	0.2	0
2	1	1	0.1	0.02	0
α = 3	1	1	0.5	0.01	600
4	1	1	0.5	0.005	1200
5	1	1	0.5	0.0025	2400
6	1	1	0.5	0.00125	4800
$r_{\alpha=3} = 0.785$					
0	1	1	1	1	0
1	1	1	0.2	0.2	0
2	1	1	0.1	0.02	0
3	1	1	0.5	0.01	0
α = 4	1	1	0.5	0.005	1600
5	1	1	0.5	0.0025	3200
6	1	1	0.5	0.00125	6400
$r_{\alpha=4} = 0.664$					

Age (x)	(1 – *survcost*) (A)	(1 – *newmort*) (B)	Survival from x – 1 to x (s_x) (C)	Age-specific survival (l_x) (l_{x-1} × A × B × C)	Age-specific fecundity (b_x)
0	1	1	1	1	0
1	1	1	0.2	0.2	0
2	1	1	0.1	0.02	0
3	1	1	0.5	0.01	0
4	1	1	0.5	0.005	0
$\alpha = 5$	1	1	0.5	0.0025	4000
6	1	1	0.5	0.00125	8000
$r_{\alpha=5} = 0.552$					

Before introducing a survival cost, note that if there is no survival cost of reproduction (*survcost* = 0) and no additional mortality (*newmort* = 0), the fitness associated with each age at maturity in Table 5.4 is the same as it was in Table 5.3.

Next, a survival cost of reproduction can be incorporated by reducing the l_x values for ages α + 1, α + 2, etc. Letting *survcost* = 0.6, reproduction reduces l_x at ages older than α (Table 5.5). After incorporating a survival cost of reproduction, we find that it has no effect on α_{opt}, but it does reduce all values of *r* for each α. This would be true of any value that we applied to the survival cost of reproduction in Table 5.5; the greater the cost, the lower the values of *r* for each age at maturity, but α_{opt} remains unchanged.

5.5.3   Incorporating new sources of extrinsic mortality

We can now explore how increases to extrinsic mortality might affect α. Extrinsic mortality is mortality resulting from factors external to the individual, such as predation, habitat alteration, climate, or disease. It is distinguished from intrinsic mortality caused by an individual's reproductive 'decisions', such as how much effort to allocate to reproduction. Intrinsic mortality is captured in the life table analyses by the parameter *survcost*.

Extrinsic mortality sources might differentially affect younger and older individuals. For example, assume that a human-induced environmental change (such as the introduction of a predatory species or urban development that reduced habitat quality for young individuals) increases extrinsic mortality during the first two years of life. This additional extrinsic mortality can be applied to the life history in Table 5.5, a life history in which fecundity and survival costs of reproduction exist. Proportional increases in extrinsic mortality at age x are reflected by *newmort$_x$*. In the revised life table (Table 5.6),

Table 5.5 *Simulated life table data for an indeterminately growing organism given in Table 5.4, expanded to accommodate a survival cost of reproduction (*survcost; *column 2) and proportional increases to natural mortality (*newmort; *column 3). Here, reproduction is assumed to reduce annual mortality after maturity such that* survcost = 0.6. *No additional extrinsic mortality is assumed, thus,* newmort = 0.

Age (x)	(1 − survcost) (A)	(1 − newmort) (B)	Survival from x −1 to x (s_x) (C)	Age-specific survival (l_x) (l_{x-1} × A × B × C)	Age-specific fecundity (b_x)
0	1	1	1	1	0
1	1	1	0.2	0.2	0
α = 2	1	1	0.1	0.02	100
3	0.4	1	0.5	0.004	200
4	0.4	1	0.5	0.0008	400
5	0.4	1	0.5	0.00016	800
6	0.4	1	0.5	0.000032	1600
$r_{α=2}$ = 0.487					
0	1	1	1	1	0
1	1	1	0.2	0.2	0
2	1	1	0.1	0.02	0
α = 3	1	1	0.5	0.01	600
4	0.4	1	0.5	0.002	1200
5	0.4	1	0.5	0.0004	2400
6	0.4	1	0.5	0.00008	4800
$r_{α=3}$ = 0.673					
0	1	1	1	1	0
1	1	1	0.2	0.2	0
2	1	1	0.1	0.02	0
3	1	1	0.5	0.01	0
α = 4	1	1	0.5	0.005	1600
5	0.4	1	0.5	0.001	3200
6	0.4	1	0.5	0.0002	6400
$r_{α=4}$ = 0.580					

Age (x)	(1 – *survcost*) (A)	(1 – *newmort*) (B)	Survival from x –1 to x (sₓ) (C)	Age-specific survival (lₓ) (lₓ₋₁ × A × B × C)	Age-specific fecundity (bₓ)
0	1	1	1	1	0
1	1	1	0.2	0.2	0
2	1	1	0.1	0.02	0
3	1	1	0.5	0.01	0
4	1	1	0.5	0.005	0
α = 5	1	1	0.5	0.0025	4000
6	0.4	1	0.5	0.0005	8000

$r_{\alpha=5} = 0.504$

newmort₁ and *newmort₂* both equal 0.50, producing a reduction in annual survival at ages 1 and 2.

Under this scenario, the optimal age at maturity has shifted slightly from age 3 to a situation in which α_{opt} could be either 3 or 4 years (Table 5.6). Note that the additional extrinsic mortality applied at ages 1 and 2 resulted in *r* being negative for individuals maturing at age 2 ($r_{\alpha=2}$ = –0.073). Recall that if *r* is negative, individuals are unable to replace themselves, meaning that a life history in which individuals matured at age 2 cannot persist through time.

Increasing the mortality at ages 1 and 2 years even further results in further increases in α_{opt} from 3 years to 4 years (when *newmort₁* and *newmort₂* = 0.55 or 0.60) and eventually 5 years when survival during the first two years of life is reduced further still (when *newmort₁* and *newmort₂* = 0.70) (Table 5.7).

5.5.4 The ratio of juvenile to adult survival

Sub-section 5.5.3 illustrated how increased extrinsic mortality during the pre-reproductive period can increase the optimal age at maturity. In the absence of new mortality from extrinsic sources (i.e. *newmort* = 0), α_{opt} was 3 years (Table 5.5). As additional extrinsic mortality steadily reduced survival at ages 1 and 2, α_{opt} had increased to age 5 (Table 5.7).

Recall that the earliest age that individuals could mature in the life tables presented in Tables 5.3–5.6 was at the end of their second year of life, i.e. age 2. This means that the increased mortality represented in Tables 5.6 and 5.7 was experienced during the pre-reproductive or pre-maturity period for every possible age at maturity.

In the study of life histories, it is important to distinguish the pre-maturity period from the post-maturity period. Ages prior to maturity comprise the juvenile stage of life; the post-maturity period comprises the adult stage of life. (These periods were initially

Table 5.6 *Simulated life table data for an indeterminately growing organism given in Table 5.5, expanded to include an increase in extrinsic mortality at ages 1 and 2 (newmort = 0.5; column 3).*

Age (x)	$(1 - survcost)$ (A)	$(1 - newmort)$ (B)	Survival from $x - 1$ to x (s_x) (C)	Age-specific survival (l_x) $(l_{x-1} \times A \times B \times C)$	Age-specific fecundity (b_x)
0	1	1	1	1	0
1	1	0.5	0.2	0.1	0
α = 2	1	0.5	0.1	0.005	100
3	0.4	1	0.5	0.001	200
4	0.4	1	0.5	0.0002	400
5	0.4	1	0.5	0.00004	800
6	0.4	1	0.5	0.000008	1600
$r_{\alpha=2} = -0.073$					
0	1	1	1	1	0
1	1	0.5	0.2	0.1	0
2	1	0.5	0.1	0.005	0
α = 3	1	1	0.5	0.0025	600
4	0.4	1	0.5	0.0005	1200
5	0.4	1	0.5	0.0001	2400
6	0.4	1	0.5	0.00002	4800
$r_{\alpha=3} = 0.256$					
0	1	1	1	1	0
1	1	0.5	0.2	0.1	0
2	1	0.5	0.1	0.005	0
3	1	1	0.5	0.0025	0
α = 4	1	1	0.5	0.00125	1600
5	0.4	1	0.5	0.00025	3200
6	0.4	1	0.5	0.00005	6400
$r_{\alpha=4} = 0.258$					
0	1	1	1	1	0
1	1	0.5	0.2	0.1	0
2	1	0.5	0.1	0.005	0

Age (x)	(1 – survcost) (A)	(1 – newmort) (B)	Survival from x – 1 to x (s$_x$) (C)	Age-specific survival (l$_x$) (l$_{x-1}$ × A × B × C)	Age-specific fecundity (b$_x$)
3	1	1	0.5	0.0025	0
4	1	1	0.5	0.00125	0
α = 5	1	1	0.5	0.000625	4000
6	0.4	1	0.5	0.000125	8000
$r_{\alpha=5}$ = 0.238					

Table 5.7 *Fitness (r) and optimal age at maturity (α$_{opt}$) associated with life histories subjected to different levels of extrinsic mortality at ages 1 and 2 (see Table 5.6). Increases in extrinsic mortality at age x are reflected by* newmort$_x$. *Values of* newmort$_1$ *and* newmort$_2$ *equal to 0.50, 0.55, 0.60, and 0.70 reflect increasing reductions in survival at ages 1 and 2, respectively.*

Reductions in annual survival at ages 1 and 2 yr reflected by newmort$_1$ and newmort$_2$	Age at maturity (α)	Fitness (r)	Optimal age at maturity (α$_{opt}$)
0.50	2	–0.073	3, 4
	3	0.256	
	4	0.258	
	5	0.238	
0.55	2	–0.150	4
	3	0.194	
	4	0.210	
	5	0.198	
0.60	2	–0.234	4
	3	0.126	
	4	0.156	
	5	0.153	
0.70	2	–0.428	5
	3	–0.037	
	4	0.026	
	5	0.044	

Table 5.8 *Life-table data extracted from Table 5.5 for individuals maturing at the end of their third year of life at $\alpha = 3$. Here, there is no additional extrinsic mortality (i.e.* newmort $= 0$*). Survival during the juvenile period is equal to* $s_0 \times s_1 \times s_2 \times s_3 = 1 \times 0.2 \times 0.1 \times 0.5 = 0.010$. *Survival during the adult period is equal to* $0.4s_4 \times 0.4s_5 \times 0.4s_6 = (0.4 \times 0.5) \times (0.4 \times 0.5) \times (0.4 \times 0.5) = 0.008$.

Age (x)	(1 − *survcost*) (A)	(1 − *newmort*) (B)	Survival from $x-1$ to x (s_x) (C)	Age-specific survival (l_x) ($l_{x-1} \times$ A $\times$ B $\times$ C)	Age-specific fecundity (b_x)
0	1	1	1	1	0
1	1	1	0.2	0.2	0
2	1	1	0.1	0.02	0
$\alpha = 3$	1	1	0.5	0.01	600
4	0.4	1	0.5	0.002	1200
5	0.4	1	0.5	0.0004	2400
6	0.4	1	0.5	0.00008	4800

Table 5.9 *Survival during the juvenile and adult periods of life for individuals maturing at age 3 associated with reductions in survival during the first two years of life. The values in column 1 correspond to values of 0, 0.1, 0.2, 0.3, ... , 0.9 for both* newmort$_1$ *and* newmort$_2$ *(see also Tables 5.4 and 5.5). The juvenile period extends from birth to age α (i.e. 3 yr in this example). The adult period extends from age $\alpha +1$ to age 6. The ratio of juvenile-to-adult survival is* juv:adult$_{survival}$.

Reduction in survival during ages 1 and 2	Survival during the juvenile period	Survival during the adult period	*juv:adult$_{survival}$*
newmort $= 0$	0.0100	0.008	1.25
newmort $= 0.1$	0.0081	0.008	1.01
newmort $= 0.2$	0.0064	0.008	0.80
newmort $= 0.3$	0.0049	0.008	0.61
newmort $= 0.4$	0.0036	0.008	0.45
newmort $= 0.5$	0.0025	0.008	0.31
newmort $= 0.6$	0.0016	0.008	0.20
newmort $= 0.7$	0.0009	0.008	0.11
newmort $= 0.8$	0.0004	0.008	0.05
newmort $= 0.9$	0.0001	0.008	0.01

distinguished in Figure 4.3 when discussing the allocation of fixed resources.) As will be discussed in Chapter 6, different regimes of juvenile and adult mortality are expected to result in the evolution of different life histories (Gadgil and Bossert 1970; Charlesworth 1980; Promislow and Harvey 1990). For example, as extrinsic mortality at potentially reproductive ages increases, selection is expected to favour those individuals (genotypes) that start to reproduce prior to those vulnerable ages, thus increasing their chances of contributing genes to future generations.

When exploring how optimal age at maturity can be affected by changes to extrinsic mortality (Tables 5.6 and 5.7), it was not the *absolute* changes in survival during the juvenile period that resulted in changes to α_{opt}, but changes in juvenile survival *relative to* survival during the adult period.

To demonstrate this, baseline survival for the juvenile and adult periods is calculated from the life-table data in Table 5.5 for individuals maturing at age 3 (Table 5.8). As explained in the caption to Table 5.8, survival during the juvenile and adult periods is 0.010 and 0.008, respectively, in the absence of additional extrinsic mortality (i.e. *newmort* = 0). Table 5.9 shows how the ratio of juvenile to adult survival, i.e. *juv:adult$_{survival}$*, changes with reductions in survival during the juvenile period; survival during the adult period remains unchanged (Table 5.9). Figure 5.3 further illustrates these changes in the ratio of juvenile to adult survival.

Although the examples in Table 5.9 and Figure 5.3 are for individuals maturing at $\alpha = 3$, it should be noted that the changes in *juv:adult$_{survival}$* with reductions in juvenile

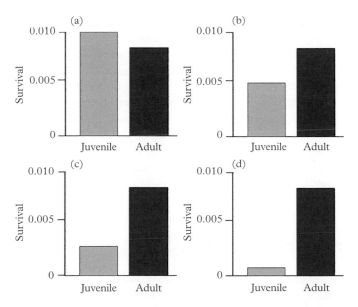

Figure 5.3 *Survival during the juvenile and adult stages for individuals maturing at age 3. Survival during the juvenile period is steadily reduced by setting* newmort *to (a) 0, (b) 0.3, (c) 0.5, and (d) 0.8. See columns 2 and 3 of Table 5.9 for the numerical values of juvenile and adult survival.*

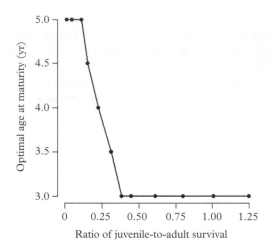

Figure 5.4 *Optimal age at maturity declines as the ratio of juvenile-to-adult survival (juv:adult$_{survival}$) increases. The figure is based on calculations of optimal age at maturity, using the information in Tables 5.5 and 5.9.*

survival are proportionately the same for individuals maturing at other ages. This can be illustrated as follows, using the data in Table 5.5 as the baseline.

As mortality during ages 1 and 2 increases from *newmort* = 0 to 0.1 and then to *newmort* = 0.2, *juv:adult$_{survival}$* changes from 1.25 to 1.01 to 0.80 when α = 3 (Table 5.9). As proportional changes, 1.25 : 1.01 : 0.80 is the same as 1 : 0.81 : 0.64. For individuals maturing at age 2, as mortality during ages 1 and 2 increases from *newmort* = 0 to 0.1 and then to *newmort* = 0.2, *juv:adult$_{survival}$* changes from 12.50 to 10.12 to 8.00. As with α = 3, these proportional changes are also 1 : 0.81 : 0.64. And for individuals maturing at age 4, as mortality during ages 1 and 2 increases from *newmort* = 0 to 0.1 and then to *newmort* = 0.2, *juv:adult$_{survival}$* changes from 0.125 to 0.101 to 0.080; simple division shows that these changes are also in the ratio of 1 : 0.81 : 0.64.

Finally, by plotting the optimal ages at maturity against *juv:adult$_{survival}$* we find that as the ratio of juvenile to adult survival increases (meaning that survival during the adult period is increasingly low relative to survival during the juvenile period), the optimal age at maturity declines (Figure 5.4).

5.6   Summing Up and a Look Ahead

Age-specific schedules of survival and fecundity comprise the vital rates that are key to estimating individual and population rates of increase. At the individual level, rates of increase reflect fitness. Common metrics include the intrinsic rate of increase (r), net reproductive rate (R_0), and reproductive value (RV). Life tables provide a straightforward means of comparing the fitness of alternative life histories, allowing for the calculation of optimal values of traits such as age at maturity. By changing parameter values, life tables can be used to explore how different levels of intrinsic (survival costs of reproduction) and extrinsic mortality influence the age at maturity that maximizes fitness.

One implication of these exploratory exercises is the realization that different regimes of juvenile and adult mortality can result in the evolution of different life histories. As one example, optimal age at maturity is expected to decrease with increases in the ratio of survival during the juvenile period of life relative to survival during the adult period of life, i.e. *juv:adult$_{survival}$*. The next chapter explores this prediction in greater theoretical and empirical detail by examining how changes to the mean and the variance of *juv:adult$_{survival}$* affect age and reproductive effort at maturity within the context of life-history trade-offs.

6

Life-History Evolution in a Changing Environment

6.1 Shifts in the Mean and Variance of Environmental Conditions

A core premise of life-history theory is that natural selection favours age-specific schedules of survival (l_x) and fecundity (b_x) that generate the highest per capita rate of increase (r) relative to potentially alternative life histories in the same population. Thus, life-history theory can be used to predict how changes to abiotic and biotic environments might influence fundamental decisions that genotypes face concerning reproduction.

A key prediction from Chapter 5 was that optimal age at maturity (α_{opt})—the value of α associated with the highest fitness relative to potentially alternative values of α—decreases with an increase in the ratio of juvenile-to-adult survival ($juv{:}adult_{survival}$). Increased $juv{:}adult_{survival}$ occurs when the environment becomes increasingly unfavourable to adults or increasingly favourable to juveniles. Either way, adults have a reduced probability of surviving the adult stage of life *relative* to surviving the juvenile period of life. This might happen because habitat critical to adult survival is negatively altered. A novel pathogen might have greater impact on older than younger individuals. Relative to small juveniles, large adults might be increasingly likely to be exploited for human consumption.

The title of Chapter 6 refers to a changing environment. These extrinsic changes can be directional in the sense that the average conditions are getting progressively better or worse for organisms. For example, the environment might change in such a way that $juv{:}adult_{survival}$ becomes higher or lower than it was previously. If this happens, we can anticipate an evolutionary change in life history. Directional change can also be manifest in terms of the variance in environmental conditions. The higher the variance, the greater the uncertainty that future generations will experience environmental conditions similar to those experienced by their parents. Thus, increased variance in $juv{:}adult_{survival}$ can also generate life-history evolution. The second half of the chapter focuses on those life histories that have evolved in response to environmental unpredictability. These are life histories that have allowed organisms to 'hedge their bets' by evolving conservative or diversification bet-hedging strategies.

A Primer of Life Histories: Ecology, Evolution, and Application. Jeffrey A. Hutchings, Oxford University Press. © Jeffrey A. Hutchings 2021.
DOI: 10.1093/oso/9780198839873.003.0006

To reiterate a key point made in Chapter 5, when studying the influence of survival on life-history evolution, it is important to separate mortality attributable to extrinsic, non-reproductive sources from that associated with intrinsic physiologically and energetically based reproductive costs. It is important because it is the extrinsically sourced components of age-specific survival that initiate or drive selection on life-history traits. Extrinsic mortality might well have consequences for the level of reproductive effort expended, thus affecting survival costs of reproduction. But in the absence of changes to extrinsic mortality, intrinsic mortality caused by survival costs are unlikely, or are far less likely, to change.

6.2   Evolution of Age and Reproductive Effort at Maturity

6.2.1   Semelparity versus iteroparity

The importance of separating juvenile from adult mortality was emphasized by Charnov and Schaffer (1973) in their examination of what is termed 'Cole's paradox'. One of the fundamental questions that Cole (1954) explored was: How many more offspring does a semelparous organism (breed once and then die) need to produce for semelparity to be favoured by natural selection over iteroparity (breed multiple times before death)? Surprisingly, he concluded that a semelparous population with mean fecundity of $(b + 1)$ would have the same rate of increase as an iteroparous population with mean fecundity b. For example, an iteroparous fish that produced one million eggs at spawning would be equally fit as one that produced one million and one eggs, and then died immediately thereafter. The paradox is this: if an extremely small increase in fecundity is all that is needed for selection to favour semelparity over iteroparity, why is semelparity so uncommon?

Although the simplest models are often the most generalizable, Cole's approach was too simple. His paradoxical result was based on a model that included neither age structure nor mortality. By distinguishing juvenile from adult mortality, Charnov and Schaffer (1973) showed that for a semelparous organism to have the same fitness as an iteroparous organism, the semelparous organism needed to produce the same number of offspring as the iteroparous organism *plus* an additional number of offspring *proportional to* the ratio of adult-to-juvenile survival. In other words, as the relative survival of adults compared to juveniles increases (i.e. a decreasing *juv:adult$_{survival}$*), thus allowing for more offspring to be produced over multiple years in an iteroparous organism, the greater the number of offspring that the semelparous organism would need to produce to have the equivalent fitness of an iteroparous organism.

Put another way, as *juv:adult$_{survival}$* increases, selection favours a semelparous over an iteroparous life history because there is a lower relative chance of survival as an adult over multiple reproductive years. And if an increase in reproductive effort increases the likelihood of meeting the additional fecundity requirements associated with adopting a semelparous life history, one can see how increased *juv:adult$_{survival}$* would also select for increased reproductive effort (given the lower chances of surviving after maturity and of reproducing more than once) (Figure 6.1).

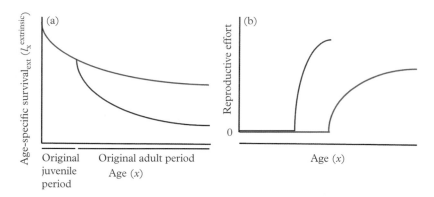

Figure 6.1 *Graphical representations of predictions of life-history theory. (a) Age-specific survival, due to extrinsic causes alone, shifts from the original curve (blue) to a new curve (red), reflecting increased adult mortality. This increase in the ratio of juvenile-to-adult survival, juv:adult$_{survival}$, is predicted to cause shifts in (b) how reproductive effort (RE) varies with age. Compared to the original pattern of RE with age (blue), the new pattern (red) reflects an earlier age at maturity, an increased RE (steeper slope), and a shorter reproductive period (i.e. an increased probability of semelparity).*

6.2.2 Life-history responses to changes in survival

As shown in Chapter 5, a key prediction of life-history theory is that increased *juv:adult$_{survival}$* favours younger age at maturity (Promislow and Harvey 1990). Another is that increased *juv:adult$_{survival}$* favours increased reproductive effort at maturity (Gadgil and Bossert 1970). Independently of mathematical simulations, these predictions about how selection should change age and reproductive effort at maturity make intuitive sense. As extrinsic mortality at potentially reproductive ages increases, selection would be expected to favour those individuals (genotypes) that start to reproduce prior to those vulnerable ages, thus increasing their probability of contributing genes to future generations.

If juvenile and adult survival both decline, but the declines are the same relative to each other, such that *juv:adult$_{survival}$* does not change, theory would not predict changes in traits such as age and reproductive effort at maturity (Figure 6.2). As noted by Gadgil and Bossert (1970: 18), 'a change in mortality does not affect the optimal reproductive effort, provided that such a change does not affect the different stages in the life history in a differential manner'. It is useful to be reminded that these predictions depend on changes in juvenile survival *relative* to adult survival. A decline, for example, in the survival of adult salmon at sea is unlikely to generate changes in life history if survival during the juvenile, freshwater phase of life declines by a proportionately similar amount.

The first, and one of the best, efforts to empirically test these predictions in wild populations was undertaken by Reznick and colleagues on guppies (*Poecilia reticulata*) that inhabit streams in Trinidad (Reznick et al. 1990). Guppies were transferred from an area in which predation on adults was high relative to that on juveniles (high *juv:adult$_{survival}$*) to a habitat where predation on adults was greatly reduced (low *juv:adult$_{survival}$*). Thirty

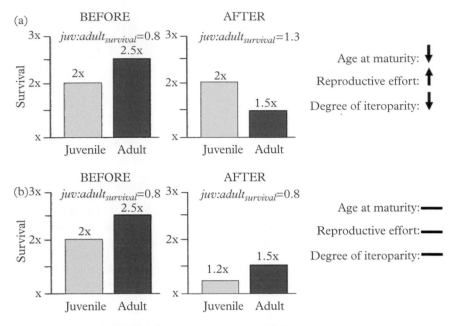

Figure 6.2 *Changes in survival will not affect life history if the changes do not differentially affect juveniles relative to adults. (a) Compared to the baseline scenario (BEFORE), an increase in juv:adult$_{survival}$ is predicted to reduce the optimal age at maturity and increase optimal reproductive effort (AFTER). These changes in life history, particularly the increased reproductive effort, would be expected to reduce the number of reproductive episodes per lifetime (lowering the degree of iteroparity). (b) Changes in survival that affect juveniles and adults to the same proportional extent, thus resulting in no change to juv:adult$_{survival}$, would not be expected to affect life history.*

to sixty generations later, the descendants of the guppy population that had been transplanted to the low *juv:adult$_{survival}$* environment 11 years earlier were found to have experienced genetically based, evolved changes in life history, maturing at an older age and allocating less of their body mass to egg production, as predicted by theory (Reznick et al. 2007).

The simulations in section 5.5 and the guppy transplant study in Trinidad represented shifts in *juv:adult$_{survival}$* within single populations. This approach allows one to predict how both α_{opt} and population viability (reflected by the *r* associated with α_{opt}) might change over time with changes to how the external environment affects survival. But this approach need not be restricted to predicting single-population responses to environmental change. It can also be used to address the question: What is the adaptive significance of life-history differences among populations of the same species?

A study of pumpkinseed sunfish (*Lepomis gibbosus*) in Ontario, Canada, took this approach. Fox and Keast (1991) compared the life histories of pumpkinseeds from five populations that experienced either high or low levels of over-winter mortality as adults. Females in the high adult-mortality environments (high *juv:adult$_{survival}$* environment)

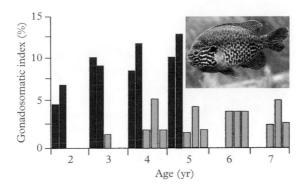

Figure 6.3 *Gonadosomatic index (mass of eggs relative to body mass) and female reproductive age in five populations of pumpkinseed sunfish. Juv:adult$_{survival}$ is high in two populations (red) and low in three others (green). Population data are presented in the same sequence at each age.*

Data from Fox and Keast (1991). Photo of pumpkinseed (slightly cropped) © Simon Pierre Barrette CC BY-SA 3.0.

matured earlier and allocated a greater proportion of their body tissue to their gonads than populations in low adult-mortality habitats (Figure 6.3).

6.2.3 Life-history responses to changes in the variance in survival

Environments differ to greater or lesser degrees across all spatial and temporal scales. From an evolutionary perspective, this raises the question of how selection would be expected to operate on the life histories of organisms subjected to unpredictable, sometimes extreme, environmental variability through time.

Murphy (1968) was the first to clearly articulate a set of predictions as to how changes to the variance (σ^2) in juvenile survival, relative to the variance in adult survival, i.e. $\sigma^2(juv:adult_{survival})$, might affect life histories. Based on mathematical models, one of his objectives was to explore the conditions under which uncertain survival for pre-reproductive individuals, but relatively stable conditions for adults, might favour itero-parity over semelparity. Supported by an admittedly sparse data set showing a positive correlation between length of reproductive period and variation in spawning success, Murphy predicted that high $\sigma^2(juv:adult_{survival})$ would favour delayed maturity, long lifespan, and multiple reproductions, but that low $\sigma^2(juv:adult_{survival})$ would select for early reproduction, high fecundity, and few reproductions, if not semelparity.

Leggett and Carscadden (1978) provided a stronger empirical anchor for Murphy's predictions. They examined life-history differences among 13 populations of American shad (*Alosa sapidissima*) throughout the species range in eastern North America. A key premise of their work was that northern populations (New Brunswick, Canada; >45° N) experienced greater variability in environmental factors important for juvenile fish survival, such as temperature, than more southerly populations (Florida to North Carolina;

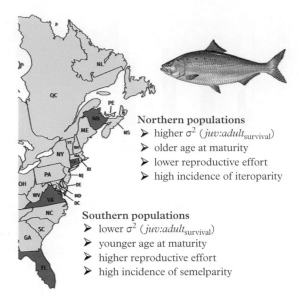

Northern populations
➤ higher σ^2 ($juv:adult_{survival}$)
➤ older age at maturity
➤ lower reproductive effort
➤ high incidence of iteroparity

Southern populations
➤ lower σ^2 ($juv:adult_{survival}$)
➤ younger age at maturity
➤ higher reproductive effort
➤ high incidence of semelparity

Figure 6.4 *Population differences in American shad life history. Green areas identify general locations of five populations studied in detail by Leggett and Carscadden (1978). From north to south, the populations were located in NB = New Brunswick (n = 2); CT = Connecticut; VA = Virginia; FL = Florida. The variance in survival during the juvenile period relative to that during the adult period is σ^2 ($juv:adult_{survival}$); reproductive effort reflects number of eggs per unit of body size.*

Drawing of American shad © Raver Duane/Wikimedia Commons.

30–35° N). In Murphy's (1968) terms, this would mean that $\sigma^2(juv:adult_{survival})$ would increase from south to north, leading to the predictions that northern populations would exhibit higher degrees of iteroparity (greater incidence of repeat spawning), older age at maturity, and lower reproductive effort (quantified as fecundity, controlling for differences in body mass). Leggett and Carscadden's (1978) findings were consistent with Murphy's predictions (Figure 6.4).

6.3   Life-History Evolution in a Variable Environment

6.3.1   Hedging evolutionary bets

At the time that Murphy (1968) was contemplating how variation in survival might affect life histories, Cohen (1966) published a paper in which he tried to explain why it was that in some semelparous species of plants not all seeds germinate immediately upon dispersal. Some fraction of the seeds remains dormant in a 'seed bank', delaying germination to subsequent years. Cohen interpreted this pattern of germination as a means by which annual, semelparous species can 'spread risk' (for perennial, iteroparous plants, risk is already spread over time).

The idea is this: When the environmental conditions for successful germination vary from one year to the next, some years being better than others, selection might favour the production of seeds with multiple germination times, allowing annual plants to spread the risk of their seeds facing unfavourable conditions over several years, i.e. several generations. It would allow them to avoid the negative fitness consequences of having all of the seeds face poor conditions in any single year. For semelparous species, a forfeit of 100 per cent of the seeds would mean the end of a plant's genotypic evolutionary line.

Cohen's (1966) model, then, was about how an annual, semelparous plant might be able to spread risk among generations. Gillespie explored a somewhat similar question, focusing on reproductive episodes within a single generation (Gillespie 1974). He was interested in how the variance in the number of successfully produced offspring might affect fitness. Simply, his idea was that if a reproductive episode can fail to produce any viable young, a parent producing, say, five offspring over its lifetime is better off by spreading those offspring over five distinct reproductive episodes rather than putting them all in one episode (Gillespie 1974). Even though expected fitness (measured as lifetime offspring output) is identical under the two strategies, spreading the risk among episodes is favoured because it reduces risk of low fitness, thus *reducing the variation in fitness* across generations. His key conclusion was that unpredictable and variable environments can generate selection that favours the production of fewer offspring in a breeding episode than the maximum that an organism is capable of producing during that episode.

In a commentary on Gillespie's (1974) paper, Slatkin (1974: 704) characterized such restraint as 'hedging one's evolutionary bets'. Since then, 'bet-hedging' has been considered a life-history strategy that reduces the fitness costs of producing offspring in occasionally unfavourable conditions at the expense of lower fitness benefits when conditions are favourable.

6.3.2   Reducing the variance in fitness

An underlying current to these considerations is that fitness is not a static entity (Figure 6.5). Environments are variable, meaning that fitness will vary through time because of stochastic, unpredictable temporal and spatial changes in quality. For illustration, fitness can be designated as W, to represent a generic, arithmetic measure of fitness (such as λ; sub-section 5.3.3) for an annual plant (meaning that each year represents a separate generation).

The mean value of fitness for a genotype through time, across generations, is termed the 'long-run fitness' (or 'long-run growth' at the population level) and can be designated here as $\bar{\bar{W}}$. A key question arises: How should $\bar{\bar{W}}$ be calculated? Should it be the arithmetic mean of W across n generations, or is there a more appropriate means of estimating long-run W that better accounts for fitness variability, such as the geometric mean (Figure 6.5)?

A simple example serves to illustrate how the arithmetic mean can inaccurately reflect long-term rates of increase. Table 6.1 summarizes population changes in abundance associated with different per-generation rates of change. The arithmetic mean of the five

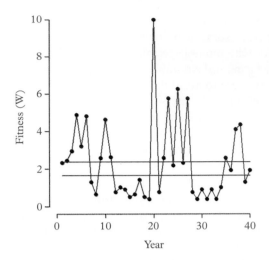

Figure 6.5 *The fitness of a genotype is not constant through time. Rather, it varies from one generation (year, in this case) to the next. Simulated data (solid circles) provide one example of how such variability might be manifest. The red line identifies the arithmetic mean of fitness (W) for the data set (2.37). The blue line is the geometric mean of W (1.63).*

Table 6.1 *Changes in population abundance across five generations.*

Generation (t)	Population size at the beginning of generation t	Population size at the end of generation t	Absolute change in population size during generation t	Per-generation rate of change %
1	10 000	11 000	1 000	10
2	11 000	5 500	−5 500	−50
3	5 500	6 325	825	15
4	6 325	7 906	1 581	25
5	7 906	8 302	396	5

per-generation rates of change (column 5 in Table 6.1) is one per cent. But a comparison of the population size at generation 5 (N = 8302) with the original population size (N = 10 000) reveals that an average one per cent per-generation increase over five generations is inaccurate. The population has declined, not increased. The actual rate of change is (8302 − 10 000) ÷ 10 000 = − 17 per cent.

This example draws attention to the fact that fitness is not so much determined by an arithmetic process as it is by a multiplicative process (Figure 6.6). The total number of descendants left by an individual after t generations depends on the product of the number surviving to reproduce in each generation (Seger and Brockmann 1987).

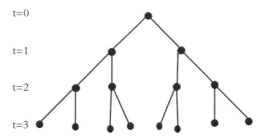

t=0

t=1

t=2

t=3

Figure 6.6 *Simulated genealogy comprised of individuals that each produces two offspring. The number of descendants left by a genotype is better described as a multiplicative process than an additive, arithmetic process. (t refers to generation.)*

Think back to the main conclusion drawn by Gillespie (1974): the advantage gained by a genotype by producing many offspring in a good year does not balance the disadvantage of producing few offspring in a bad year. This is what the example shown in Table 6.1 also illustrates. Four generations of positive rates of change were insufficient to offset a single bad year of negative change.

Thinking of fitness as a multiplicative process allows for the use of the geometric mean to calculate long-run fitness, $\bar{W}$. Rather than basing the mean on a summation of a set of values, as the arithmetic mean does, the geometric mean is based on the product of a set of values. After t generations, the geometric mean of W can be calculated as:

$$\bar{W} = [(W_1) \times (W_2) \times (W_3) \times (W_4) \times \ldots \times (W_t)]^{(1/t)}$$

Equation 6.1

or

$$\bar{W} = \sqrt[t]{W_1 \times W_2 \times W_3 \times W_4 \times \ldots \times W_t}$$

Equation 6.2

or

$$\bar{W} = e^{\text{mean}(\log(W))}$$

Equation 6.3

A comparison of arithmetic and geometric means for a population growing over ten generations is provided in Table 6.2. After ten generations, the population size according to the arithmetic mean of λ is $N_{10} = N_1 \times \lambda^{10} = 100 \times 1.80^{10} = 35\ 705$. However, according to the geometric mean of λ, after ten generations, $N_{10} = N_1 \times \lambda^{10} = 100 \times 1.64^{10} = 14\ 075$, which is within rounding error (1.6 per cent) of the actual value of 13 849.

6.3.3 Bet-hedging life histories

There are two useful things to remember about the geometric mean. Firstly, it is strongly influenced by unusually low values. Secondly, the more variable a set of values, the lower the geometric mean. If environmental unpredictability contributes to high variability in fitness, thus resulting in a low geometric mean $\bar{W}$, we would expect natural selection to

Table 6.2 *A population growing for ten generations. The initial abundance is 100 individuals. The rate of increase shown is* λ *(which is equal to N_{t+1}/N_t). The number of new individuals each generation is given by* λN_t.

Generation (t)	N_t	λ	$N_{t+1} = \lambda\, N_t$
1	100	1.05	105
2	105	1.65	173
3	173	1.11	192
4	192	2.72	522
5	522	3.32	1733
6	1733	1.82	3154
7	3154	1.01	3186
8	3186	2.86	9112
9	9112	1.49	13 577
10	13 577	1.02	13 849
Arithmetic mean		**1.80**	
Geometric mean		**1.64**	

act to reduce the variance in fitness across generations and increase the geometric mean. That is, we would expect natural selection to favour a bet-hedging life-history strategy.

There are two types of bet-hedging strategies. The first is termed a *conservative* bet-hedging strategy. Conservative bet-hedging is characterized by a reduction in risk; it represents an insurance policy against bad times. A conservative bet-hedger might, for example, reduce the probability of having poor reproductive episodes unduly influence fitness by switching to dormant eggs earlier than would be expected in a season as a safe, conservative hedge against the likelihood of, say, an early frost. Traits exhibited by conservative bet-hedgers include delayed maturity, larger body size, lower reproductive effort, greater longevity, and increased breeding events per lifetime (Table 6.3).

The black-browed albatross (*Thalassarche melanophrys*), a potential conservative bet-hedger, matures at ~10 yr and lives to 44 yr (Myhrvold et al. 2015). Among birds, it is one of the oldest at maturity (Figure 2.3) and among the longest lived (Figure 2.7). Two populations differ three-fold in the variability of sea-surface temperatures adjacent to the breeding colonies. The South Georgia population, exposed to a high-variability environment, experiences higher adult survival and produces fewer hatchlings per capita than those on Kerguelen (Figure 6.7). To unequivocally be an example of conservative bet-hedging, and not simply different fitness optima across environments, it would be necessary to confirm that the conservative traits expressed by birds on South Georgia do

Table 6.3 *Examples of traits associated with conservative and diversification bet-hedging life-history strategies. The traits listed here are not intended to be mutually exclusive. Organismal examples are from Simons (2011).*

Strategy	Bet-hedging trait	Examples
Conservative	↓ fruit or flower to ovule ratio	Angiosperms
	↓ reproductive effort	Reptiles, birds, trematodes
	Diapause	Copepods, arthropods
	↓ clutch size	Birds
	↑ age at maturity	Angiosperms, birds, fishes
	↑ size at maturity	Birds, fishes, mammals
	↑ gestation period	Reptiles
	↑ lifespan	Fishes, mammals
	↑ degree of iteroparity	Fishes, mammals
Diversification	Variable offspring size	Plants, fishes, amphibians, arthropods, bryozoans, molluscs, arachnids, gastropods, annelids, polychaetes, echinoderms, echinoids, ascidians
	Variable development time	Amphibians, arthropods
	Egg hatching asynchrony	Birds, branchiopods, amphibians, nematodes
	Polyandry	Amphibians, arthropods, gastropods
	Polygyny	Mammals
	Variable germination time	Angiosperms
	Variability in egg or larval diapause	Arthropods, branchiopods, rotifers
	Variable dormancy responses	Angiosperms, branchiopods, tardigrades

not maximize expected fitness within each generation but do maximize long-term fitness across generations.

Atlantic cod provides another potential example of a conservative bet-hedger. Females delay maturity until a large size has been attained. Larger individuals produce greater numbers of eggs and spawn more times within a longer spawning period. Although earlier maturation and highly synchronous spawning may increase the expected fitness within a generation, the conservative traits decrease the variance in fitness across generations by increasing the chances that offspring will hatch when food is available.

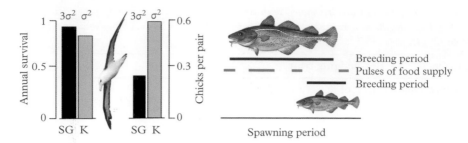

Figure 6.7 *Postulated conservative bet-hedging species. Black-browed albatross on South Georgia (SG) experience temperatures three times more variable (3σ²) than those on Kerguelen (K) Islands. Across all adult breeding types, SG albatrosses experience higher annual survival and expend lower apparent reproductive effort (Nevoux et al. 2010). Atlantic cod delay maturity, allowing for higher numbers of reproductive episodes and longer spawning period than smaller cod, increasing the likelihood that larvae will hatch in the presence of food (Hutchings and Rangeley 2011).*

Black-browed albatross photo © J.J. Harrison (CC BY-SA 3.0). Line drawing of Atlantic cod © H.L. Todd.

The second strategy is *diversification* bet-hedging. It is generated by selection for increased phenotypic variation within genotypes (such as the same plant producing seeds of different sizes; Capinera 1979). By producing a diverse array of phenotypes among offspring, the probability of having poor reproductive episodes unduly affect fitness is reduced, thereby increasing the chances that some of one's offspring will experience favourable environments.

Traits exhibited by diversification bet-hedgers include polyphenism (two or more distinct phenotypes are produced by the same genotype), polyandry/polygyny, and variability in seed/egg size, germination time, diapause, and development time (Table 6.3). In addition to these metrics of variability, Simons (2007) makes the important point that diversification bet-hedging can also generate selection for increased fecundity independently of selection for offspring size. Using simulations, he showed that higher offspring number leads to a higher geometric mean under environmental uncertainty.

Many insects and plants exhibit diversification bet-hedging (Table 6.3). Polyphenism in pea aphids (*Acyrthosiphon pisum*) is one example. Parthenogenetic females produce genetically identical winged and wingless daughters; winged offspring are advantageous when conditions are crowded, wingless offspring when conditions are not (Figure 6.8). Variability in the timing of seed germination, evident even under controlled environmental conditions, provides another example of diversification bet-hedging. Working with Indian tobacco (*Lobelia inflata*), Simons and Johnston (2006) found that variation in germination time differs among genotypes within the same population. This suggests that selection can act on the *variance* in germination timing, a finding consistent with the hypothesis that such diversification bet-hedging is an adaptive, evolved strategy (Simons 2009).

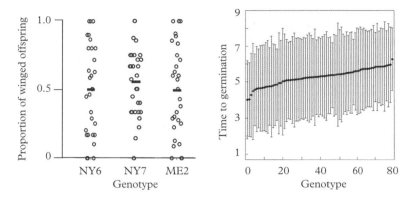

Figure 6.8 *Diversification bet-hedging species. Left panel: Three genotypes of parthenogenetic pea aphids produce offspring who themselves produce offspring of widely differing proportions of winged and wingless females; black bars represent means (Grantham et al. 2016). Right panel: The within-genotype variance in standardized germination time (see Simons and Johnston (2006) for details), represented by the vertical lines (95% confidence intervals), differs among* Lobelia inflata *under controlled conditions. Black diamonds represent means.*

Left panel reproduced by permission from the Royal Society. Right panel reproduced by permission from John Wiley and Sons.

6.3.4 Deterministic versus stochastic *r*

A key theme of this chapter is that the fitness of a genotype varies through time. Much of this variability can be attributed to various sources of stochastic, or unpredictable, variation (Lande 1993). *Genetic* stochasticity is caused by random changes in allele frequencies over time. It is manifest by genetic drift and is likely to have its greatest effect on fitness when populations are quite small, when the probability of inbreeding is high (just how small populations need to be before genetic stochasticity is of primary importance is a matter of some debate). *Demographic* stochasticity, also thought to be of increasing importance as populations decline, is reflected by random changes in l_x and b_x for some, but not all, individuals in a population. *Environmental* stochasticity similarly affects l_x and b_x for all individuals in a population and can be of importance when populations are either large or small. Individual fitness, and correspondingly population viability, can also be affected by *catastrophic* environmental perturbations.

An excellent treatment of how stochasticity affects long-run fitness (defined as $\bar{r}$) was provided by Lande (1993) and expanded upon by Sæther and Engen (2015). Based on their work, and that of others, suffice to say that stochastically based estimates of $\bar{r}$ are consistently less than deterministic estimates. Stochastic estimates incorporate variability in key determinants of fitness, such as individual survival, fecundity, and growth. Deterministic estimates are based on the assumption that age-specific rates of survival, fecundity, and growth are constant. The greater the stochasticity in l_x and b_x, the lower that stochastically based estimates of $\bar{r}$ will be when compared to deterministic estimates of $\bar{r}$.

Thus far, we have been estimating fitness in a deterministic manner (e.g. Tables 5.1–5.6). Although the assumption of constancy in life-history traits represents a simplifying assumption, it does not mean that analyses based on deterministic estimates of r are somehow misleading. The appropriateness of estimating r, using deterministic versus stochastic methods, depends on the question being asked. For example, if you are interested in predicting qualitative or directional changes in life-history traits as a result of environmental perturbation, deterministic estimates will generally be fine. But if you want to estimate the probability that specific qualitative or directional changes in traits will occur, a stochastic modelling framework would be preferable.

To illustrate how stochasticity can be incorporated in estimates of r, we first need to define the frequency distribution, or parameter space, for age-specific rates of survival and fecundity. Ideally, this variability would be based on data obtained from the population of interest. When such data are unavailable, these distributions might be based on information for conspecific populations or closely related species. Clearly, the stronger the empirical basis for the frequency distributions of l_x and b_x, the greater the confidence we can have in the analytical results.

Consider the estimation of fitness associated with a semelparous life history for which individuals reproduce at 3 years of age (and, of course, die shortly thereafter). Based on data obtained from the same population (ideally), fecundity at age 3 (b_3) is normally distributed at age x; the mean and standard deviation of the distribution are 33.00 and 7.95, respectively (Figure 6.9). Although a normal distribution often fits fecundity data well, a different distribution is required to model variability in survival. The rationale for selecting a beta distribution to model survival data is based on its simplicity, smoothness, and flexibility, making it an ideal choice for distributions that have restricted support (i.e. whose range is limited), in this case between 0 and 1 (Figure 6.9). For our population of interest, the mean value of l_3 is 0.038.

Although probability distributions of the intrinsic rate of increase cannot be described by a stochastic model (because of the analytical constraint that r can only be calculated by iteration), r can be approximated by the natural logarithm of the net reproductive rate

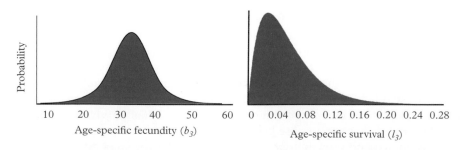

Figure 6.9 *Age-specific fecundity and survival for a semelparous organism breeding at age 3. The normal (left) and beta (right) distributions are used in a stochastic model to estimate r.*

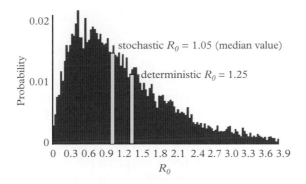

Figure 6.10 *Frequency distribution of net reproductive rate, R_0, as determined from 10 000 iterations of a stochastic model for a semelparous organism for which $l_3 = 0.038$ and $b_3 = 33$.*

(R_0) discounted by generation time, as discussed in section 5.4. Recall that the equation for the net reproductive rate (Equation 5.1) is:

$$R_0 = \sum_0^\infty l_x b_x$$

For the stochastic modelling, this equation for R_0 will be 'run' 10 000 times. During each simulation run, a value of b_3 will be 'drawn' from the normal distribution in Figure 6.9 and a value of l_3 will be drawn from the beta distribution. Given that the distribution of b_3 is normal, with a mean (μ) of 33 and a standard deviation (σ) of 7.95, there is a statistical probability of 68.3 per cent that the value of b_3 drawn from the distribution will be between 25.05 and 40.95, i.e. $\mu \pm \sigma$. Correspondingly, there is a very low but non-zero probability that the value of b_3 drawn is less than 20 or greater than 50.

The end result of this exercise will be 10 000 values of R_0. The frequency distribution of these values extends from a lower limit of zero (R_0 cannot be negative), reaching peak probabilities (albeit still very small) of between 0.3 and 0.9, before exhibiting an extended tail at larger values (Figure 6.10). The stochastic value of R_0 is lower than the deterministic value. Converting these values of net reproductive rate ($r \sim \ln(R_0)/G$; see section 5.4) yields stochastic and deterministic values of r of 0.016 and 0.074, respectively.

6.4 Summing Up and a Look Ahead

The underlying current to this chapter is environmental change. Environments are not static. They can shift directionally or exhibit natural variability, sometimes predictably (e.g. seasonal periodicity) but often unpredictably (stochasticity). Environmental change that affects age-specific rates of survival and fecundity will affect the evolution of life-history traits. Examples of life-history evolution in response to environmental change include predictable responses to selection in traits such as age at maturity and reproductive effort. These predictions are based on shifts in the ratio of juvenile-to-adult survival and shifts in the variance of juvenile survival relative to the variance in adult survival.

Variability is the key word for the second half of the chapter. In response to environmental unpredictability, organisms have evolved bet-hedging strategies that maximize the geometric mean, or long-run fitness. These life histories can involve one or more conservative or diversification bet-hedging traits. Under semelparity, selection can favour the germination of seeds or the hatching of diapausing eggs across multiple generations. Under iteroparity, rather than producing the maximum number of offspring that an organism is capable of producing in few breeding episodes, environmental variability can favour the production of fewer offspring per episode but across a greater number of breeding episodes. This trade-off has implications for the evolution of offspring size and offspring number, among the most wide-ranging of life-history traits, discussed next in Chapter 7.

7

Number and Size of Offspring

7.1 Extreme Variability in the Production of Propagules

Thus far, this book has discussed variability in the number and size of offspring from two perspectives. The first, illustrated in Chapter 2, documented the tremendous differences that exist in offspring size and number across taxa. The smallest seed (1 µg in orchids) and the largest whale calf at birth (2250 kg in blue whales) differ in mass by 10^{12}. Number of offspring per individual per breeding event is similarly variable, ranging from one or two (many birds, mammals, and reptiles) to several hundred million in marine fishes such as ocean sunfish (*Mola mola*), greasy grouper (*Epinephelus tauvina*), and black marlin (*Makaira indica*). The second topic that encompassed offspring size and number was bet-hedging (sub-section 6.3.3). Conservative bet-hedging can involve the production of fewer offspring than what an individual is capable of producing. One potential means (but certainly not the only mechanism; Table 6.3) of achieving diversification bet-hedging is manifest by differences in the size of propagules produced by a single individual.

Whenever tackling questions across a phylogenetic breadth of any magnitude, it is wise to first explore the degree to which differences among species are the result of different evolutionary histories. For example, differences in offspring size:number combinations can be attributed to a phylogenetic constraint, a result or component of the phylogenetic history of a lineage that prevents an anticipated course of evolution in that lineage (McKitrick 1993). As discussed earlier (sub-section 2.3.1), some combinations of traits are less likely because of genetic, developmental, physiological, and/or structural differences associated with a species' evolutionary history. Once we account for trait variation generated by constraints, we can then turn to the question of whether differences in offspring number and size might be attributable to selection. In this regard, two approaches have dominated the life-history literature.

The first concerned offspring number in a specific group of organisms. The fundamental question: Why don't birds produce more eggs per clutch than the modest numbers observed under natural conditions, especially when they seem capable of doing so? Put another way: What are the factors that influence natural selection for offspring number? For those who have explored this question (and there have been many), initially

A Primer of Life Histories: Ecology, Evolution, and Application. Jeffrey A. Hutchings, Oxford University Press. © Jeffrey A. Hutchings 2021.
DOI: 10.1093/oso/9780198839873.003.0007

in birds and then in other small-clutch organisms, such as mammals and lizards, the focus tended to be on number rather than size of offspring.

The second approach had its origins in optimality modelling. The primary interest was not offspring number. Rather, the focus was on how natural selection acts on the 'investment per offspring' that maximizes fitness. Given that investment per offspring can potentially encompass many elements of reproductive effort (such as nutritional provisioning of each seed/egg, feeding of individuals post-hatch, energetic costs of parental care), the concept has potential to be broadly applicable among species beyond those that produce relatively few offspring per breeding event. For many researchers, especially those working on fishes and plants, the optimality approach to understanding the evolution of propagule size has been of greater interest than understanding selection for propagule number.

These two approaches differ in a conceptually fundamental way. The clutch-size approach focuses on how natural selection acts on numbers of offspring per breeding episode; the size of offspring is generally of secondary interest. If offspring size is considered, it is within the context of how this trait can change as a result of parental nourishment and care. In contrast, the conceptual point of departure of the second, investment-per-offspring approach is that selection acts primarily on offspring size and that the number of offspring produced is a by-product of this selection process.

Before venturing further, a quick word about terminology. The term used to describe the number of propagules produced by an individual during a single breeding episode depends on the taxonomic group under study. Clutch size tends to be favoured by those studying birds and insects; litter size by mammal researchers; fecundity by those studying fishes; and seed number or count by plant biologists.

7.2   Offspring Number and Size: Not All Options Are Possible

Intuitively, all else being equal, we can imagine selection favouring the production of large offspring and many of them. The larger the offspring, presumably the better its provisioning and the greater its opportunity for survival during the precarious early stages of life. The more offspring produced, the greater the chance that some will live to reproduce successfully themselves.

Before considering how offspring number and size are affected by selection, it is useful to be reminded that these traits can be phylogenetically constrained to differ from what we might otherwise expect. Phylogenetic constraints tend to be most obvious when the phylogenetic lineages under comparison are very distantly related. Consider the amniotes, vertebrates which produce eggs that have an amnion (a membrane that encompasses the embryo when it is first formed). The amnion fills with amniotic fluid, serving to protect the developing embryo. Amniotic eggs allow for efficient exchange of gases and wastes between the developing embryo and the atmosphere. The development of an outer shell may have allowed for larger eggs to be produced, compared to the egg sizes characteristic of fishes and amphibians (anamniotes). Mammals, and some sharks

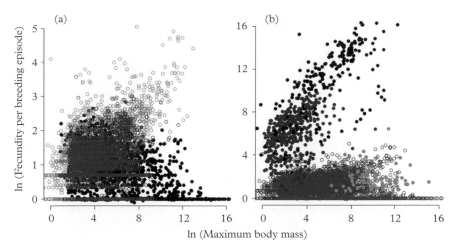

Figure 7.1 *Natural-log-transformed fecundity (i.e. clutch/litter size) per breeding episode as a function of log-transformed maximum body mass (grams). Panel (a) shows the data for amniotes: 6625 species of birds (blue), 3420 species of mammals (black), and 1790 species of reptiles (green). In panel (b), these 11 835 species of amniotes (blue) have lower fecundities than 6676 species of amphibians (red) and 209 species of bony fishes (black). Data on 82 chondrichthyan fishes (sharks, skates, rays; green, panel (b)) reveal fecundities similar to those of amniotes. Data from Myhrvold et al. (2015) for amniotes, Oliveira et al. (2017) for amphibians, and Barneche et al. (2018) and Hutchings et al. (2012) for fishes.*

and rays, that do not lay eggs develop corresponding structures for protection, gas exchange, and waste removal in the uterus.

The amniotes include birds, reptiles, and mammals. A bivariate plot of maximum fecundity against maximum body size (plotted on a log-log scale to improve clarity; Figure 7.1(a)) reveals that mammals attain the largest sizes among amniotes but that reptiles produce the highest number of offspring per clutch. Birds are intermediate. The average number of offspring per clutch/litter ($\pm$ se) for the three classes of amniotes is 3.16 ± 0.02 for birds, 7.89 ± 0.28 for reptiles, and 2.56 ± 0.03 for mammals.

Inclusion of data for other vertebrate classes makes it quite clear that most anamniotes produce far more offspring per breeding episode than amniotes (Figure 7.1(b)). The average for amniotes (3.70 ± 0.05) is orders of magnitude less than that for amphibians (1034 ± 73) and bony fishes ($1\,535\,471 \pm 408\,405$). This is likely indicative of a phylogenetic constraint. Birds, mammals, reptiles, and chondrichthyan fishes are prevented from producing several millions of offspring per individual because of developmental constraints imposed by the production of comparatively large eggs that are laid on land or by fertilized embryos that are retained and nourished within a restricted uterine space in the mother. Selection for increased number of propagules in amniotes appears to be constrained by a phylogenetically induced limit of about 150 (5 on the natural log scale in Figure 7.1(a)), the maximum recorded clutch size of snakes in the family Viperidae.

Across the range of potential clutch sizes, perhaps the most unexpected number is one. These species provide a single opportunity per clutch for parents to provision or

1 offspring
per clutch

Figure 7.2 *Species that produce one offspring per clutch or breeding episode. Little brown bat* (Myotis lucifugus); *Atlantic puffin* (Fratercula arctica); *bowhead whale* (Balaena mysticetus); *stream anole* (Anolis oxylophus).

raise an offspring until they are sufficiently well developed to obtain resources themselves. Despite what seems like a risky strategy, a surprising number of species produce only one offspring per clutch or litter (Figure 7.2). Based on a dataset of >23 000 species of reptiles, birds, and mammals (Myhrvold et al. 2015), almost 1 400 species have a clutch size of one: 763 mammals, 475 birds, and 132 reptiles. Given that the combined number of bird, mammal, and reptile species is ~25 000, this means that the incidence of a single-offspring per clutch strategy is roughly five per cent among amniotes. Some examples include most bats (~1 400 species; order Chiroptera), puffins (three species; *Fratercula* spp.), many cetaceans (~90 species), and anoline lizards (~425 species; *Anolis* spp.).

Reptiles provide ample evidence of constraints on offspring number. The order Squamata is the second most speciose order of vertebrates (>10 000 species). Some squamates produce exceedingly few eggs per clutch: geckos, two per clutch; anoline lizards, one per clutch. It has been estimated that such small clutch sizes have evolved independently more than 20 times in lizards (Shine and Greer 1991). In addition to constraints imposed by phylogeny, clutch size in squamate reptiles may be constrained by mechanical and locomotory constraints associated with arboreality (living in trees), narrow crevices, and foraging mode (summarized by Roff 1992). In anoles, egg size is constrained by the aperture of the pelvic girdle (Michaud and Echternacht 1995).

7.3   Evolution of Offspring Number

7.3.1   Early thinking: clutch size in birds

Among the earliest attempts to understand life-history variability in wild populations were questions related to the clutch size of birds. As early as the 1830s, natural historians had noted a latitudinal pattern in clutch size (Rensch 1938): birds in tropical regions

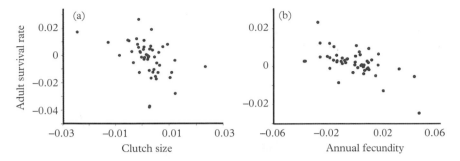

Figure 7.3 *Scatterplots of negative associations between adult survival and (a) clutch size and (b) annual fecundity among bird families and orders, controlling for phylogeny.*

Source: Bennett and Owens (2002). Reprinted by permission from Oxford University Press.

tend to have smaller clutches than those in more northerly regions. Interestingly, the earliest explanations for this pattern also incorporated geographical differences in juvenile and adult mortality as a causal basis for the latitudinal trend.

Championed by German ornithologists Erwin Stresemann (whose major study spanned 1927 to 1934) and Bernhard Rensch (1938), the argument was that birds in more northerly regions produced larger clutches as an adaptive response to higher mortality rates (Klomp 1970). As Moreau (1944: 300) hypothesized, 'it is possible to suggest a series of interactions between clutch-size and total mortality by which each tends to replace the other'. In other words, increases in clutch size might represent a response to offset the losses generated by increased mortality, or reduced survival. The prescience of Stresemann, Rensch, and Moreau was much later realized by Bennett and Owens' (2002) negative associations between adult survival and fecundity in birds (Figure 7.3).

Notwithstanding early attempts to understand inter-specific variability in offspring number within an ecological context, it was the work of David Lack, beginning in the late 1940s and extending into the 1960s, that was to have the greatest influence on the thinking of how clutch size evolved in endotherms (and later insects). Although Lack was first and foremost a field ecologist, his thinking was in evolutionary terms, asserting that litter size is a trait under natural selection and that 'the genotypes selected are those which result in the maximum number of descendants' (Lack 1948: 46).

7.3.2 The Lack clutch size

Lack put the problem in this way (Lack 1947a). Consider a bird species that normally lays four eggs per clutch. Why should it not normally lay five eggs? He reasoned implicitly that there must be a negative correlation between clutch size (x) and the average survival of offspring (y). The linear function in Figure 7.4 (red line) illustrates a simple form of such a relation. If offspring survival declines with increasing clutch size, it raises the question as to what might be causing this decline. Lack concluded that clutch size is

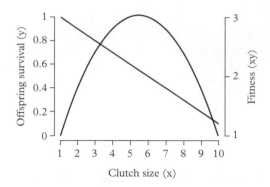

Figure 7.4 *Lack's clutch-size model. As clutch size (x) increases, average offspring survival (y) (red line) decreases. Parental fitness (xy) is maximized at an intermediate clutch size (black curve). Here, the most productive clutch, or 'Lack clutch size', is five to six eggs.*

primarily influenced by the number of offspring that parents can feed and raise to the fledgling stage.

The fitness of a parent can be calculated as the product of clutch size x and the average survival of offspring (y) in a clutch of size x, yielding (in the present example) the dome-shaped relationship depicted in Figure 7.4. The most productive clutch size, often termed the 'Lack clutch size', is that yielding the highest parental fitness for that clutch.

7.3.3   Observed clutch sizes often differ from the Lack clutch size

Despite its tremendous influence, deficiencies in Lack's approach have been well-documented: (i) the model ignores adult survival; (ii) the most productive clutch is unlikely to be the same throughout life; (iii) eggs and their incubation are not cost-free; (iv) factors other than parental feeding ability almost certainly affect clutch size.

The interesting thing is that Lack himself was aware of most of these issues. He just didn't think they were particularly important determinants of clutch size in birds. Regarding adult survival, despite Lack's citing of Moreau's (1944) work, and the latter's emphasis on the influence of juvenile and adult mortality on avian clutch size, he discounted the idea that increased clutch size had a 'weakening effect on the parent of laying too many eggs', arguing that this 'can safely be ruled out, since clutch-size is so far below the potential limit of egg-production' (Lack 1947b: 21). (Egg production and incubation in birds do have costs; Monaghan and Nager 1997.) He also discounted the notion that larger broods might have lower average survival because of increased conspicuousness to predators, at least for grey partridge (*Perdix perdix*) in England (Lack 1947b).

A tremendous amount of research has been devoted to testing Lack's ideas about clutch size. A dominant impression that many of these later papers leave in the reader's mind is that Lack was conceptually narrow in his thinking. But a close reading of his early work reveals instead a laudable breadth.

Foreshadowing a plethora of studies finding that observed clutch sizes are smaller than the most productive (i.e. the Lack clutch size), Lack (1948) provided the following reasons for why data on guinea pigs (*Cavia porcellus*) revealed the most common litter size to be smaller than the most productive: (i) mortality after weaning (analogous to fledging in birds) is not accounted for; (ii) not all differences in litter size represent adaptation; and, rather interestingly within the context of reproductive costs, (iii) large litters might not be observed because they reduce the breeding life of the parent. He also acknowledges the potential importance of factors other than food supply on litter size, such as seasonality, predation on offspring, parental age, body size (mentioning fishes), and density (mentioning insects). The odd thing is that Lack did not generally extend his thinking on other vertebrates to his thinking of clutch-size evolution in birds.

Indeed, insects present some challenges in interpreting the adaptive significance of clutch size (a topic succinctly and informatively summarized by Roff 1992). Most species lay their eggs in, on, or near an appropriate food source or 'host'. A key trade-off for insects relates to clutch size and host availability or density. A large clutch might lead to over-crowding and intense competition among offspring for limited resources. But if a smaller clutch size necessitates a greater number of clutches, this would lead to a greater number of suitable hosts that must be found, leading to longer search times by the parent, increased vulnerability to predation, and higher adult mortality between egg-laying events. In some insects, there can be a lower limit to clutch size such that a minimum number of eggs must be laid by the female for any of them to survive. Such an aggregative effect (an example of an Allee effect; Figure 1.7) might be necessary to overcome a host's defence system.

Thus, selection on insect clutch size is thought to be a function of both number and size of offspring (Parker and Begon 1986). The same is true of fishes and plants.

7.4 Evolution of Offspring Size

7.4.1 Early thinking: a trade-off between size and number of offspring

From a general life-history perspective, perhaps the greatest deficiency in Lack's clutch-size model is the exclusion of consequences to adults, the most striking omission being a meaningful consideration of adult survival. Interestingly, it was a lack of consideration of juvenile survival that crippled Cole's paradox regarding semelparity (sub-section 6.2.1). (This serves as a useful contemporary reminder that life-history studies that ignore mortality, on either juveniles or adults, should be interpreted with a very great deal of caution.)

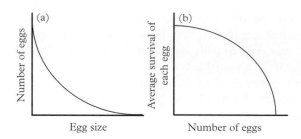

Figure 7.5 *Svärdson (1949) reasoned that upper limits to fecundity were determined by how offspring survival and egg size change with egg number per female. Two key arguments are presented graphically in (a), illustrating a convex trade-off between egg number and egg size (volume), and by (b), a concave function between offspring survival and fecundity.*

A second element largely absent in the Lack clutch-size model and its later derivatives is a consideration of offspring size.

Gunnar Svärdson appears to have been the first to explore the evolutionary implications of the trade-off between number and size of offspring. Rather than birds and mammals, the Swedish biologist was thinking of fishes (although he also published widely on birds) whose fecundity vastly exceeds that of endotherms. Svärdson (1949) suggested that there must be an upper limit to fecundity which depends on the influence of egg size on offspring survival and parental reproductive success. Otherwise, he argued, directional selection—or as he put it, a tendency to increase egg number every generation—would favour continual increases in the numbers of eggs per female. He remarked, 'From a theoretical point of view it thus is rather easy to conclude that there must also be a selection pressure for *decreasing* [his italics] egg numbers, but it is not so extremely evident how this selection works' (Svärdson 1949: 116).

Even if he was not entirely clear on how selection might operate, Svärdson (1949) was clear in setting out his arguments as to why the evolution of offspring number should be considered jointly with the evolution of offspring size. He made three assertions. Firstly, there is a negative correlation within individual females between egg number and egg size. Secondly, larger offspring hatch from larger eggs. Thirdly, larger offspring survive better than smaller offspring.

There were two key elements to Svärdson's perspective. The more important was that there is a trade-off between the number of eggs produced by a female and the size of each of those eggs. Given that his focus was on fishes, each female has a constrained gonadal volume (mm³) within which eggs develop. Assuming that egg volume is approximately that of a sphere (volume = 4/3 × π (radius in mm)³), the trade-off is convex (Figure 7.5(a)). The second key assumption was that the survival probability of recently hatched individuals declines as the number of eggs increases (Figure 7.5(b)). Of various forms that this relationship might take, a concave pattern seems defensible (although other functions are clearly possible).

Svärdson did not use graphs or equations to illustrate his verbal arguments. This might have prevented him from taking the logical step of estimating the parental fitness associated with different egg sizes by, for example, multiplying the number of eggs associated with a particular egg size (calculated from Figure 7.5(a)) by the average survival probability of each of those eggs (from Figure 7.5(b)), as Smith and Fretwell (1974) did twenty-five years later (sub-section 7.4.3).

7.4.2   Investment per offspring

Despite differences that distinguish Lack's ideas concerning birds and mammals from Svärdson's thoughts on fishes, both identified a trade-off between number and survival of offspring as being integral to limiting the total number of offspring produced per breeding episode. For Lack, the key factor limiting selection for increased clutch size was the cost associated with parental ability to provide food during the period of parental care. For Svärdson, selection for increased egg number was limited by costs to both parents and offspring associated with producing smaller eggs. Both factors pertain to the concept of parental investment per offspring.

If the reproductive effort a parent invests in its offspring is constrained, an increase in numbers of offspring will necessarily come at a cost of reduced parental investment per offspring. This investment can be in the form of the size of seed or egg, or the quality and quantity of parental care. By focusing discussion on investment per offspring, rather than the specifics of avian clutch size or the fecundity of fishes, the adaptive significance of alternative offspring size:number strategies can be extended to multiple taxonomic groups.

7.4.3   Smith–Fretwell model of optimal size and number of offspring

The concept of maximizing the rate of return on investment per offspring has intuitive appeal from an evolutionary perspective. Christopher Smith and Stephen Fretwell capitalized on this idea by proposing a simple graphical model in 1974. Their reasoning can be summarized as follows.

Assume that fitness can be approximated by the rate of return on investment per offspring (Figure 7.6). Think of the 'return' as offspring survival. Investment represents some measurable form of reproductive effort such as individual propagule size or amount of food fed per young. The rate of gain in offspring survival (y) per unit of investment per offspring (x) can be represented by the slope of a straight line, i.e. $\Delta y/\Delta x$. The straight line is anchored at the origin (zero investment yields zero return). We can imagine there being different fitness lines or fitness functions with high, medium, and low slopes corresponding to high, medium, and low fitness (dashed lines in Figure 7.6(a)).

These fitness functions do not reflect linear relationships between investment per offspring and offspring survival. Rather, each of the dashed lines simply represents a different potential slope, i.e. different potential fitness. The important thing is that realized fitness functions must overlap with the relationship between offspring survival and investment per offspring for the species or population of interest. For example,

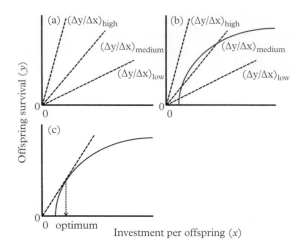

Figure 7.6 *Developing the model of optimal investment per offspring. (a) Three fitness functions, each represented by different slopes of the rate of change in offspring survival (y) per unit of change in investment per offspring (x). (b) The fitness functions in (a) relative to a red curve relating offspring survival to investment per offspring. (c) The fitness function that is tangential to the red curve identifies the optimal investment per offspring.*

Figure 7.6(b) reveals that both the medium and low fitness functions overlap such a curve relating offspring survival (x) to investment per offspring (y). The curve can be thought of as representing the set of possible combinations of x and y for a specific population. However, the medium and low fitness functions in Figure 7.6(b) do not yield the maximum slope corresponding to the set of possible combinations. It is the fitness function tangential to the curve that yields the maximum slope. The investment per offspring where the tangential fitness function touches the curve is the optimum (Figure 7.6(c)).

The Smith-Fretwell model in Figure 7.6(c) predicts optimal offspring size (investment per offspring) based on a metric of parental fitness that accounts only for offspring survival (there is also an implicit but not unreasonable assumption that the greater the survival during the offspring stage, the greater the survival throughout the rest of life). To account for the constraint that larger offspring (particularly beneficial from the offspring's perspective) are produced at a cost of producing fewer offspring (not necessarily beneficial from the parent's perspective), a function relating offspring number to offspring size is required. In species for which parental investment in offspring ends with the extrusion of an egg (e.g. most fishes, amphibians), this is a reasonably straightforward calculation (as noted in relation to Figure 7.5(a)).

Thus, the Smith-Fretwell model can be used to estimate the optimal offspring size that maximizes parental fitness within a single population. The key element to estimating

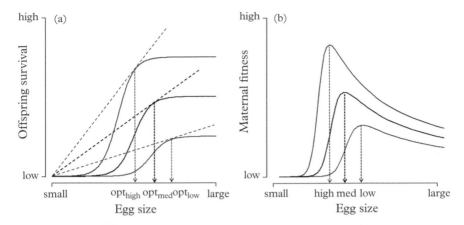

Figure 7.7 *Environmental quality is predicted to influence selection for egg size in fishes and amphibians. The relative food levels are high (blue), medium (black), and low (red). Low-quality environments, reflected perhaps by low food supply, favour the production of fewer, larger eggs; correspondingly, high-quality environments favour more numerous, smaller eggs. This hypothesis would yield offspring survival curves similar to those shown in panel (a). The maternal fitness functions in panel (b) are produced by multiplying the number of eggs of size x by the offspring survival probability for eggs of size x. The optimal egg sizes in panels (a) and (b) are indicated by vertical arrows.*

this optimum is the relationship between offspring size and offspring survival (the red curve in Figure 7.6(c)). If this relationship changes, the optimum will change. For example, optimal offspring size might differ among populations because of ecological factors that affect survival, such as food availability, competition, predation, habitat, parental care, and dispersal. For offspring size optima to differ under the Smith-Fretwell model, there must be differences in the relationship between offspring survival and investment per offspring.

Consider the effects of environmental quality on optimal offspring size. Theory and empirical work find that optimal egg size in fishes and amphibians increases as environmental quality declines (Rollinson and Hutchings 2013). This would imply Smith-Fretwell curves of the shapes shown in Figure 7.7(a). For organisms that have roughly spherical eggs, the relationship between egg size and egg number can be calculated as the gonadal volume (the space in the body cavity available for eggs) divided by the volume of each egg. This yields a function with the same shape as that in Figure 7.5(a). Maternal fitness for each environment can then be approximated by the product of offspring survival and number (holding gonadal volume constant) (Figure 7.7(b)). These results illustrate how the Smith-Fretwell model can be applied to explore the consequences to optimal egg size and maternal fitness associated with changes in environmental quality. The model predicts that optimal egg size increases as environmental quality declines (Figure 7.7(a)).

7.4.4   Selection for increased fecundity

Figure 7.7 illustrates how offspring survival and maternal fitness vary with offspring size given different levels of environmental quality. Whatever the shapes of these curves, they share the basic property of size-dependent offspring survival. For the examples illustrated in Figure 7.7, offspring survival varies continuously with egg size. Any factor expected to increase offspring survival across all egg sizes, such as food supply, is predicted to result in a reduction in optimal egg size, thus favouring females that produce relatively numerous, relatively small offspring. But it is unclear how common such a continuous relationship between offspring size and survival might be.

An obvious alternative is that offspring survival is largely independent of propagule size. In other words, the probability that an offspring survives does not depend on whether the offspring is relatively large or relatively small. This might be common in species that provide no parental care, for which offspring survival depends entirely on vagaries of the environment. Examples would include wind-dispersing plants, broadcast-spawning invertebrates (e.g. sea urchin, *Strongylocentrotus* spp., abalone, *Haliotis* spp.), and many fishes. If the likelihood of death of such a dispersed seed or egg does not decline with increasing propagule size, selection would favour individuals that maximize the production of offspring near their physiologically or developmentally minimum size. Under these circumstances, optimal egg size would be unaffected by changes in environmental quality (Figure 7.8).

Albeit at the species level (often not ideal when exploring life-history evolution, within-population data being preferable), the most comprehensive data relating propagule size to survival are for plants. Moles and Westoby (2006) found no correlation between seed mass and seed survival prior to dispersal ($n = 346$ species, based on the global

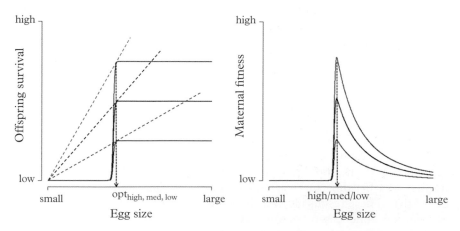

Figure 7.8 *Above a minimum propagule (egg) size, offspring survival is likely to be independent of propagule size for many species. Under these circumstances, optimal offspring size would not vary with environmental quality. The line colours are those specified in the caption to Figure 7.7.*

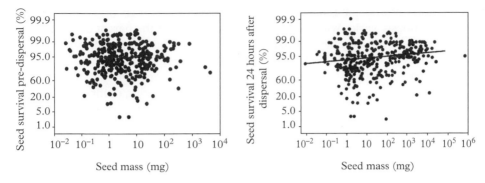

Figure 7.9 *Relationships between seed mass and seed survival before (left panel) and 24 hours after dispersal (right panel).*

From Moles and Westoby (2006). Reprinted by permission from John Wiley and Sons.

literature) (Figure 7.9). Survival during the first 24 hours following dispersal was statistically related to seed mass ($p = 0.002$; $n = 361$ species), although the data are rather widely scattered and the correlation is not significant at all spatial scales (Moles et al. 2003).

Unlike the circumstances in which offspring survival is positively related to offspring size (Figure 7.7), natural selection under size-independent survival (Figures 7.8 and 7.9) would favour a strategy that maximized the number of offspring each of which approached the physiological minimum size (below which survival declines to zero).

7.5   Summing Up and a Look Ahead

Offspring number and size are two of the most variable life-history traits. Fecundity per female ranges from a single offspring per clutch in many amniotes to hundreds of millions in some fishes and plants. The mass of the smallest seed and the largest whale calf differ by twelve orders of magnitude. Among species, much of this variability can be attributed to genetic, developmental, physiological, or structural constraints. Some trait combinations are not possible because of differences associated with a species' evolutionary history. Substantial variation in propagule number and size can exist among populations of the same species, generating questions concerning the adaptive significance of this variability. Hypotheses for the evolution of offspring number and size have been very much influenced by the study organism. Those who study species that produce relatively few offspring have tended to be drawn to theories of clutch or litter size; those who study species that produce comparatively numerous offspring have tended to be drawn to offspring-size theory.

The most influential models are those attributed to Lack on clutch size and to Smith and Fretwell on offspring size. Fundamental to both sets of models is a trade-off between offspring number and parental investment per offspring. When offspring survival or fitness continuously varies with offspring size, the fitness of the parent depends on both offspring size and the number of offspring of that size that the parent can produce. If offspring survival is independent of offspring size, parental fitness is maximized when individuals maximize the production of minimally sized propagules.

Body size can be one of the greatest determinants of offspring number and offspring size. It is also one of the primary correlates of alternative life-history strategies and tactics examined in Chapter 8.

8

Alternative Life Histories

8.1 Alternative Reproductive Phenotypes

In the 1960s, increased interest in the adaptive significance of phenotypic variability brought changes in how different behaviours among individuals in the same population were interpreted. Previously there was a strong sense that individuals should behave in a conventional or typical manner. Deviations from the perceived norm were considered sub-optimal, resulting in reduced reproductive success and lower fitness. Krebs and Davies (1981: 221) put it memorably: 'A decade ago, if an animal was seen behaving in a different way from the majority of the population it was often thought to be abnormal. Male ducks that engaged in forced copulations instead of courting females by displays were said to be behaving abnormally due to overcrowding. If we observed a male bullfrog sitting silently in the middle of a chorus, while other males were croaking loudly to attract females, we would perhaps have thought that it was ill or having a rest.'

Readily observable and easily measurable differences in behaviour are conveniently tractable attributes when studying alternative forms of mating and reproduction. Some individuals overtly fight to secure access to a mate, while others avoid fighting, finding ways to 'sneak' copulations. One male might aggressively guard a female to maintain her within his harem, not knowing that the 'female' is actually a male, poised to surreptitiously obtain a mate. Not surprisingly, these alternative behaviours have generated tremendous interest among ecologists and ethologists keen to interpret within-population differences in behaviour in an evolutionary context. (Curiously, the word 'alternative' is often used to identify phenotypes other than the most frequently observed phenotype which is presumed to be the 'conventional', 'prevailing', or 'typical' phenotype. This subjectivity is problematic, particularly when the proportionately dominant phenotype switches within a population through time or naturally differs among populations. In this chapter, the word 'alternative' will not carry connotations with respect to the frequency of the phenotype within a population.)

Although much of the early literature referred to alternative behaviours as alternative life histories, the focus tended to be on behaviour rather than life history, insofar as the life-history consequences of different combinations of behavioural traits were often unexplored. However, once the effects on l_x and b_x began to be addressed, questions

A Primer of Life Histories: Ecology, Evolution, and Application. Jeffrey A. Hutchings, Oxford University Press. © Jeffrey A. Hutchings 2021.
DOI: 10.1093/oso/9780198839873.003.0008

concerning the fitness implications and causal mechanisms of alternative behavioural phenotypes began to be addressed.

Following a broad overview across animal taxa, this chapter focuses on the conditional nature of alternative mating phenotypes. One aspect of this conditional nature is the idea that thresholds (behavioural, developmental, morphological) exist that, once attained or exceeded by an individual, trigger an alternative mating or reproductive response. There is now considerable evidence that alternative behaviours are partially genetically determined. Genetic polymorphisms (differences in allele frequencies) can generate alternative mating phenotypes, as can threshold norms of reaction, leading to the inference (and demonstration) that selection can alter the incidence or frequency of alternative behavioural phenotypes within populations. After considering the role of negative frequency-dependent selection on the evolutionary stability of alternative mating phenotypes, the chapter closes by addressing some terminological inconsistencies in the literature.

8.2   Dichotomies in Sex, Size, and Status

Alternative behaviours represent discontinuous variation at a given point in time. Many of these phenotypic discontinuities are associated with dichotomies that have common elements in most animals.

The first of these reflects the observation that males are more likely to express alternative reproductive phenotypes than females (Shuster and Wade 2003; Oliveira et al. 2008). At its core, the reason for this is that female mating success is generally limited by the number of gametes she can produce, whereas males are limited by the number of mates they can procure (Shuster and Wade 2003). Thus, female gametes and, by extension, females, constitute a limiting resource. This relative scarcity has potential to generate intense competition among males for access to females. Male–male competition, in turn, leads to selection, resulting in heritable changes in male phenotypes to increase their probability of securing mates.

Given that females are more likely to constitute a limiting resource than males, it is logical to assume that the probability of obtaining a mate will be higher for females than for males. This will result in greater variability in mating success among males; some will have considerable reproductive success, others will have little or none. All else being equal, the greater the variation in reproductive success (a reflection of fitness), (i) the greater the opportunity for, and the strength of, selection (Crow 1958; Wade and Arnold 1980), and (ii) the greater the opportunity for selection to act on existing phenotypic variation in ways that favour non-conventional or alternative mating phenotypes. It is this greater opportunity for selection, resulting from male–male competition for mates, that is largely responsible for differences between the sexes in the strength of sexual selection and a greater preponderance of alternative phenotypes among males (Shuster and Wade 2003).

There is also the question of how female choice influences the evolution of alternative male mating phenotypes (Alonzo 2008). If female fitness (perhaps through effects on survival, condition, or care of offspring) is differentially affected by the alternative male phenotype with whom she mates, we would expect these differential effects to lead to the evolution of female choice among alternative male phenotypes. However, general predictions on the outcomes of this co-evolution between the sexes are currently lacking, reflecting a need for research on how male–male competition and female choice interact to influence the evolution of alternative mating phenotypes (Alonzo 2008).

In addition to sex, alternative reproductive phenotypes conform with other apparent dichotomies (Table 8.1): territorial/aggressive vs non-territorial/passive; large vs small;

Table 8.1 *The most common correlates of alternative reproductive phenotypes in various groups of organisms, based on information in Oliveira et al. (2008).*

Organism	Alpha	Beta
Insects	Territorial or guarding female(s)	Search for females over long distances
	Aggressive (fights other males)	Sneak copulations or fertilizations
	Guard female(s)	Satellite of territorial/advertising male
Crustacea	Guard female(s)	Sneaker
	Change sex (hermaphroditism)	Usurper (replaces guarding male)
	Large size (both sexes)	Female mimicry
Fishes	Territorial	Sneaker
	Aggressive	Satellite
	Large size	Female mimicry
Amphibians	Caller (vocalizing male)	Satellite (silent, near caller males)
	Amplexus	Female mimic
	Active searcher	Spermatophore capper
Reptiles	Aggressive	Sneaker
	Territorial	Female mimicry
	Colour differences	Satellite
Birds	Territorial	Parasitism of nests by egg laying
	Aggressive	Extra-pair copulation (primarily males)
	Plumage differences	Female mimicry; satellites
Mammals	Guard female(s)	Sneaker, satellite
	Behavioural dominance/aggression	Female mimicry
	Territorial	Nomad; wanderer; searcher

advertisement vs discreetness. In species that express two alternative phenotypes, one is often classified as 'alpha' (other terms include dominant, bourgeois, conventional), the other 'beta' (subordinate, parasitic, or, more often than not, simply 'alternative'). Alpha males are typically described as being larger, territorial, aggressive, and/or overt in attracting the attention of potential mates. Alpha males obtain fertilizations by defending access to females and guarding resources necessary for reproduction, whereas beta males commonly obtain fertilizations by sneaking, by adopting a satellite or searching behaviour, or by mimicking females in colour and/or behaviour.

As illustrated by Table 8.1, alternative mating and reproductive phenotypes are taxonomically widespread. Based on the conclusions of behavioural experts in their respective taxonomic fields (Oliveira et al. 2008), alternative life histories have been documented in at least 184 families (Table 8.2).

Table 8.2 *Families in which alternative reproductive phenotypes have been documented (collated from information in Oliveira et al. 2008 and updated to reflect taxonomic changes).*

Class	Family	
Insects	Acrididae (grasshoppers)	Halictidae (sweat bees)
(*n* = 58)	Andrenidae (short-tongued bees)	Ichneumonidae
	Anostostomatidae (wetas)	Libellulidae (dragonflies)
	Anthophoridae (digging bees)	Lucanidae (atlas beetles)
	Apidae (social bees)	Lycaenidae (butterflies, moths)
	Bethylidae	Lygaeidae (chinch bugs)
	Bittacidae (hanging flies)	Megachilidae (leafcutter bees)
	Braconidae	Megapodagrionidae (damselflies)
	Brentidae	Meloidae (blister beetles)
	Bruchidae (seed beetles)	Neriidae (cactus flies)
	Calopterygidae (damselflies)	Nymphalidae (butterflies, moths)
	Cerambycidae (longhorn beetles)	Panorpidae (scorpionflies)
	Ceratopogonidae (blood-sucking midges)	Phlaeothripidae (thrips)
	Chalcididae	Phoridae
	Chironomidae (midges)	Pneumoridae (bladder grasshoppers)
	Coelopidae (seaweed flies)	Pompilidae (spider wasps)
	Coenagrionidae (damselflies)	Pteromalidae
	Colletidae	Rhopalidae (soapberry bugs)

Class	Family	
	Ctenophthalmidae (fleas)	Satyridae (butterflies, moths)
	Delphacidae (planthoppers)	Scarabaeidae (dung beetles)
	Drosophilidae (fruitflies)	Scatophagidae (dung flies)
	Dryomyzidae	Silphidae (burying beetles)
	Empididae (dancing flies)	Sphecidae (digger wasps)
	Forficulidae (earwigs)	Staphylinidae (rove beetles)
	Formicidae (ants)	Syrphidae (hoverflies)
	Oestridae (botflies)	Tenebrionidae (fungus beetles)
	Gerridae (water striders)	Tephritidae (true fruitflies)
	Gryllidae (crickets)	Tettigoniidae (katydids)
	Gryllotalpidae (mole crickets)	Vespidae (wasps)
Crustaceans	Alpheidae (snapping shrimp)	Ocypodidae (crabs)
($n = 32$)	Aoridae (amphipods)	Oregoniidae (crabs)
	Chirocephalidae	Palaemonidae (shrimp)
	Crangonidae (shrimp)	Palinuridae (spiny lobster)
	Dotillidae (crabs)	Pandalidae (shrimp)
	Epialtidae (crabs)	Portunidae (crab)
	Gnathiidae (isopods)	Processidae (shrimp)
	Gonodactylidae (mantis shrimp)	Pseudosquillidae (mantis shrimp)
	Grapsidae (crabs)	Rhynchocinetidae (shrimp)
	Idoteidae (isopods)	Sicyoniidae (prawn)
	Inachidae (crabs)	Sphaeromatidae (isopod)
	Ischyroceridae (amphipods)	Tachidiidae (copepod)
	Janiridae (isopods)	Talitridae (isopod)
	Limnadiidae (branchiopods)	Tanaididae (isopod)
	Lysmatidae (shrimp)	Thoridae (shrimp)
	Nephropidae (lobster)	Triopsidae (tadpole shrimp)
Fishes	Acanthuridae (surgeonfish)	Labridae (wrasses)
($n = 33$)	Acipenseridae (sturgeons)	Macroramphosidae (snipefish)
	Adrianichthyidae (medaka)	Mochokidae (upside-down catfish)

(Continued)

Table 8.2   Continued

Class	Family	
	Batrachoididae (toadfish)	Monacanthidae (filefish)
	Blenniidae (combtooth blennies)	Osphronemidae (gouramies, fighting fish)
	Catostomidae (suckers)	Ostraciidae (boxfish)
	Centrarchidae (sunfishes)	Percidae (perch)
	Chaetodontidae (butterflyfishes)	Pinguipedidae (sandperch)
	Cichlidae (cichlids)	Poeciliidae (livebearers)
	Cyprinidae (minnows, carps, loaches)	Polycentridae (leaffish)
	Cyprinodontidae (pupfish)	Pomacentridae (damselfish)
	Esocidae (pike)	Salmonidae (salmon, trout, char)
	Gadidae (cod)	Scaridae (parrotfish)
	Gasterosteidae (sticklebacks)	Serranidae (sea bass)
	Gobiidae (gobies)	Sparidae (porgies)
	Hexagrammidae (greenlings)	Trypterygiidae (triplefin blennies)
	Hypoptychidae (sand eels)	
Amphibians	Ambystomatidae (mole salamanders)	Leptodactylidae (ditch frogs, dwarf frogs)
(*n* = 12)	Bufonidae (toads)	Myobatrachidae (froglets, toadlets)
	Cryptobranchidae (giant salamanders)	Plethodontidae (lungless salamanders)
	Eleutherodactylidae (rain frogs)	Ranidae (true frogs)
	Hylidae (frogs: cricket, tree, chorus)	Rhacophoridae (frogs: foam-nest, flying)
	Hynobiidae (Asian salamander)	Salamandridae (true salamanders, newts)
Reptiles	Agamidae (dragon lizards)	Lacertidae (lacerta lizards)
(*n* = 12)	Chelydridae (snapping turtles)	Phrynosomatidae (spiny, horned lizards)
	Colubridae (snakes)	Pythonidae (pythons)
	Crocodylidae (crocodiles)	Scincidae (skinks)
	Emydidae (pond turtles)	Teiidae (lizards: whiptails, racerunners)
	Gekkonidae (gekkos)	Typhlopidae (blind snakes)
Birds	Accipitridae (buzzards, hawks, eagles)	Muscicapidae (Old World flycatchers)

Class	Family	
(*n* = 11)	Cardinalidae (buntings, grosbeaks)	Passerellidae (New World sparrows)
	Cuculidae (cuckoos)	Passeridae (Old World sparrows)
	Estrildidae (pipits, seedcrackers)	Scolopacidae (ruff, sandpipers)
	Fringillidae (finches)	Stercorariidae (skuas, jaegers)
	Monarchidae (paradise flycatchers)	
Mammals	Antilocapridae (pronghorns)	Felidae (cats)
(*n* = 26)	Atelidae (red howlers)	Galagidae (bushbabies)
	Balaenopteridae (baleen whales)	Hominidae (gorilla, chimpanzee, orang-utan)
	Bovidae (bison, antelopes, sheep)	Indriidae (sifaka)
	Callitrichidae (marmosets, tamarins)	Lemuridae (lemurs)
	Cebidae (capuchins, squirrel monkeys)	Lorisidae (pottos)
	Cercopithecidae (macaques, baboons)	Macropodidae (wallabies, kangaroos)
	Cervidae (deer)	Mustelidae (otters, weasels)
	Cheirogaleidae (dwarf lemurs)	Phocidae (seals)
	Cricetidae (voles, lemmings)	Phyllostomidae (bats)
	Elephantidae (elephants)	Physeteridae (sperm whales)
	Emballonuridae (microbats)	Procyonidae (raccoons)
	Equidae (horses)	Sciuridae (squirrels, marmots)

8.3   Thresholds and Conditional Tactics

Much of the literature on alternative mating phenotypes has focused on their 'conditional' nature. The idea is that the probability that an individual will express an alpha or beta phenotype depends, or is conditional, on some aspect of their intrinsic (physiological, hormonal) or extrinsic environment, including the phenotype or behaviour of others with whom they are interacting.

If a favourable set of intrinsic or extrinsic conditions exists for switching phenotypes, the establishment of these favourable conditions can be considered analogous to the attainment of a 'threshold'. If we imagine a normally distributed variable that is

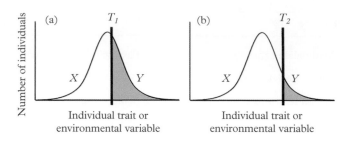

Figure 8.1 *Thresholds (T_1, T_2) for intrinsic individual traits or extrinsic environmental variables determine the proportion of individuals in a population that adopt alternative phenotypes X and Y. In panel (a) roughly equal proportions of the population adopt the alternative phenotypes. In panel (b) the incidence of phenotype Y is considerably reduced.*

representative of condition, such as depicted in Figure 8.1, the position of a threshold identifies the proportion of the population that, on average, adopts each of the alternative phenotypes. In Figure 8.1(a), the threshold T_1 is such that 45 per cent of individuals have exceeded it and will express phenotype Y rather than the alternative phenotype X. The threshold in Figure 8.1(b) is such that 15 per cent of individuals have surpassed T_2 and will express phenotype Y.

Thresholds need not have a genetic basis, allowing them to change throughout life. They might not reflect developmentally constrained pathways, thus allowing individuals to rapidly switch between alternative phenotypes. Many behaviourally governed thresholds are likely to be based on and change with past learning and experience. This seems manifest in the caller-satellite dichotomy in anurans (frogs and toads); callers use vocal communications to attract females while silent male satellites 'parasitize' fertilizations (Zamudio and Chan 2008). Individuals often switch between the calling and satellite tactics within a single night.

Alternative mating or reproductive phenotypes (particularly behaviours) that are not genetically based and that are readily reversible are often referred to as alternative 'tactics' (a discussion of the terminological morass that envelopes this literature is offered in section 8.7). For example, in the absence of relatively larger males, an individual might be territorial. But in the presence of males larger than himself, the same individual might instead adopt a submissive tactic and await opportunities to sneak copulations. Under these circumstances, although the conditional 'rule' for adopting a specific tactic might be genetic (e.g. fight if larger, flee if smaller), the underlying tactical threshold—the body size at which the switch occurs—need not be.

Alternative tactics need not afford equal fitness. The probability of attaining some thresholds, such as those based on body size, may be random with respect to genotype. Individuals fated by the environment to adopt a sub-optimal phenotype have lower fitness as they make the best of a bad situation (better to mate with reduced probability of success than to not reproduce at all).

For completeness, it should be noted that a lack of genetic variability in thresholds could also be caused by genetic monomorphism, meaning that all individuals in a population share exactly the same threshold. Under these rather improbable circumstances, there would be no heritable variation in thresholds among individuals and the thresholds would not respond to selection.

8.4   Genetic Polymorphisms

If the absence of an underlying genetic basis represents one end of a continuum of influences on alternative phenotypes, the other extreme is genetic polymorphism. Indeed, there are some species for which the expression of a particular phenotype depends entirely on genetic architecture. Genetically determined phenotypes are not reversible within an individual's lifetime. Because of their long-term, developmentally constrained nature, genetic polymorphisms are often termed 'strategies'. Examples follow for species of crustaceans, fishes, lizards, and birds.

A crustacean for which alternative strategies have been widely documented is the marine isopod *Paracerceis sculpta* (Figure 8.2). Native to intertidal and subtidal zones in the eastern Pacific, *P. sculpta* breed inside spongocoels (the large central cavity of sponges) of the calcareous sponge *Leucetta losangelensis*. The species has been well-studied in the northern Gulf of California where males are characterized by three discrete strategies (females are monomorphic) (Shuster and Wade 1991, 2003; Shuster 2008). In terms of size, alpha (α) males are the largest, beta (β) are intermediate, and gamma (γ) males are smallest. α-males defend harems within sponges; β-males invade harems by mimicking female behaviour; γ-males invade harems by being small and secretive. These alternative strategies are determined by allelic variation at a single locus and by interactions with alleles at other loci.

In addition to sex ratio, fertilization success in *P. sculpta* appears to be frequency dependent, meaning that the fitness of a strategy depends on its incidence relative to other strategies in the population. If only one of the three morphs is present in a

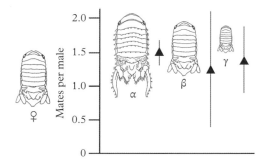

Figure 8.2 *Average number of mates (±95% confidence interval) for α, β, and γ male strategies in the isopod* Paracerceis sculpta.

Redrawn from Shuster and Wade (2003). Line drawings © Stephen Shuster (CC BY 3.0). Image altered to reposition the female to the y-axis.

Figure 8.3 *Male* Neolamprologus multifasciatus *interacting by a* Neothauma tanganyicense *shell in the wild (photo credit Jakob Guebel) (Bose et al. 2020).*

Reprinted by permission from the Royal Society.

spongocoel, they sire all offspring. If more than one male morph is present, reproductive success depends on both the number of females and on the relative frequencies of the male morphs (Shuster and Wade 1991). Averaging across morphs, there are no differences in the number of mates obtained by males of each type (Figure 8.2), an observation consistent with the hypothesis that they have equal fitness.

Different tactics can be expressed within genetically polymorphic strategies. Shell-dwelling cichlid fishes endemic to East Africa's Lake Tanganyika are obligate snail brooders. Two examples are *Neolamprologus multifasciatus* (Bose et al. 2020) and *Lamprologus callipterus* (Taborsky 2008). Females spawn in empty shells, preferably those of the snail *Neothauma tanganyicense*, after which they remain in the shell and guard the eggs until they hatch (Figure 8.3).

Male *L. callipterus* express two alternative strategies. A Mendelian polymorphism distinguishes the larger-morph strategy from the dwarf strategy (Taborsky 2008). Larger-morph males exhibit two tactics: nesting and sneaker. Nesting males collect empty shells and defend them, attracting a harem of usually two to six females. During egg deposition, nesting males release sperm through the opening of the shell at the same time that sperm is released by competing males adopting the alternative sneaker tactic. Dwarf males (~2.5 per cent the mass of nesting males) enter shells with spawning females, squeeze themselves into the shell's inner whorl to avoid detection, and attempt to fertilize eggs from their advantageous position.

The nesting and sneaker tactics of the larger-morph strategy are reversible and appear to be conditional on body size. If the number of empty shells available for nesting is limited, adoption of the nesting tactic depends on whether males are large enough to

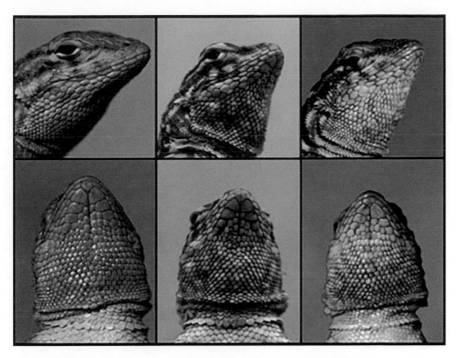

Figure 8.4 *Alternative strategies in male side-blotched lizards* (Uta stansburiana). *These annual lizards develop an orange, blue, or yellow throat at maturity (six to eight months).*

Source: *Sinervo and Lively (1996). Reprinted by permission from Springer Nature.*

transport shells into their nest. If empty shells are widely available, males do not need to carry and accumulate them, allowing the nesting tactic to be expressed at lower body-size thresholds. The observation that sneakers are generally smaller than nesting males provides additional evidence that the tactics are size based.

Although the existence of genetically polymorphic alternative strategies in both males and females has not commonly been reported, the most widely studied species in this regard is likely the side-blotched lizard, *Uta stansburiana* (Sinervo and Lively 1996; Sinervo et al. 2000).

Three male morphs are distinguished by throat colour (Figure 8.4). Orange-throated males are highly aggressive, establish large territories, and defend access to large groups of females. Blue-throated males are far less aggressive, defend small territories, and tend towards social monogamy. Yellow-throated males are furtive, non-territorial, mimic females when confronted by territorial males, and sneak fertilizations on the territories of other males. The three male morphs are the product of a genetic polymorphism comprised of a single locus with three alleles (Zamudio and Sinervo 2000; Sinervo and Zamudio 2001). Behaviourally, the orange-throated morph is dominant over the blue-throated morph from which it can steal mates. The blue-throated morph, usually having only one female to defend, can successfully guard against sneak copulations by the yellow-throated morph. The latter, however, aided by its ability to mimic females, can successfully sneak fertilizations in the large territories of orange-throated males.

The three-way relationship of dominance and subordinance creates a cyclical pattern of frequencies of the three morphs over time. When rare, competition with other members adopting the same strategy is low, as is the variance in male reproductive success. A morph can increase in frequency until its fitness starts to decline because of increasing intra-morph competition (and increasing variance in reproductive success), after which its frequency declines while that of an alternative morph increases (Sinervo and Lively 1996). As with the isopod *P. sculpta*, the genetic polymorphism in side-blotched lizards is an example of a species in which alternative strategies appear to be maintained by frequency-dependent selection (discussed in section 8.6), such that the fitness of each strategy depends on its incidence in the population.

Section 8.2 drew attention to the fact that alternative strategies can exist in females. In the side-blotched lizard, yellow-throated females produce relatively few, large eggs whereas orange-throated females produce greater numbers of smaller eggs (Sinervo et al. 2000; Alonzo 2008). The fitness associated with these strategies is thought to be linked with population density. When density is low, orange-throated females are favoured because of their greater offspring numbers; when density is high, yellow-throated females are favoured because of the higher survival of their larger offspring. The fitness and frequency of female morphs oscillate with population density in a two-year cycle (Sinervo et al. 2000). Yellow- and orange-throated females are also able to exert some control over which male fertilizes their eggs, suggestive of differences in their patterns of mate choice (Calsbeek and Sinervo 2002). Thus, the fitness of alternative strategies in male side-blotched lizards is very likely affected not only by the frequencies of the male strategies but by female choice as well (Alonzo and Sinervo 2001).

A fourth example of how a genetic polymorphism can generate alternative reproductive strategies goes beyond allelic variability at single genes to reveal an integral role of genetic architecture. Central to the mating system of the ruff (*Philomachus pugnax*), a Eurasian sandpiper, is mate competition among three male morphs (Lank et al. 1995; Küpper et al. 2016) (Figure 8.5). 'Independent' males, distinguished by dark plumage, aggressively defend territories and attempt to exclude other males from the breeding lek. Slightly smaller 'satellite' males have white colouration, are less aggressive, non-territorial, and co-display along with the independents. Non-territorial 'faeder' males, smaller than independents and satellites but larger than females, are female mimics (Jukema and Piersma 2006). Their plumage is that of a female, they do not engage in displays, and they are often courted by displaying males. They are not perceived by independent and satellite males to represent a competitive threat, allowing faeders to maintain close proximity to females (Küpper et al. 2016).

The differences in plumage and behaviour among the male morphs are controlled by coadapted gene complexes located within large, 4.5 Mb (4 500 000 base pairs) inversions on chromosome 11 (Küpper et al. 2016; Lamichhaney et al. 2016). An inversion is a chromosomal rearrangement in which portions of DNA are 'flipped' or reversed end to end (Rieseberg et al. 2001). Importantly, inversions suppress recombination, allowing

Figure 8.5 *Male ruffs exhibit three alternative strategies. 'Independents' have dark ruffs and head tufts and aggressively defend territories. 'Satellites' have white ruffs/head tufts, are non-territorial, and co-display with independents. Faeders are female mimics. The strategies are determined by the presence/absence of supergene variants on chromosome 11. The section of the chromosome containing the inversion is indicated by the coloured arrows. The ancestral independents lack the supergene. The Faeder variant shows the inversion (red) originating 3.8 MYA. The supergene of the Satellite variant shows a mixture of red and blue, representing a recombination event 0.52 MYA, followed by subsequent accumulation of genetic change, as estimated by Lamichhaney et al. (2016). The inversions are present on only one chromosome; homozygosity for the inversion is lethal.*

Photograph of ruffs © Susan McRae, used with permission.

the multiple genes they contain to be inherited as single linked units, or 'supergenes' (there are ~ 125 genes in the ruff supergene; Küpper et al. 2016). The independents are considered the ancestral strategy because they lack the inversion. Of the two variants to the ruff chromosomal inversion, the Faeder variant is estimated to have originated 3.8 million years ago and the satellite variant through a rare recombination event 520,000 years ago (Lamichhaney et al. 2016).

Supergenes are known to underlie complex phenotypes in an increasing number of species (Schwander et al. 2014; Oomen et al. 2020). In birds, in addition to the ruff, alternative strategies in the white-throated sparrow (*Zonotrichia albicollis*) are also controlled by supergenes contained within a chromosomal inversion. The tan morph (both sexes) does not possess the inversion whereas the white morph (both sexes) contains a ~ 100 Mb supergene (Tuttle et al. 2016). Drawing upon an example in plants, a chromosomal inversion polymorphism is responsible for the existence of two adaptive life-history ecotypes of the yellow monkey flower, *Mimulus guttatus* (Lowry and Willis 2010; Twyford and Friedman 2015).

8.5   Genotype-by-Environment Thresholds for Alternative Strategies

8.5.1   Threshold reaction norms

As interest in their evolutionary significance grew, dimorphic phenotypes were increasingly thought of as threshold traits, i.e. characters determined by alleles at multiple loci that can be assigned to one of two or more distinct classes (Roff 1996). The framing of alternative strategies within the context of threshold traits has had numerous conceptual and practical advantages (as illustrated in this sub-section and those following), not the least of which is that they are readily amenable to genetic analysis. By 1950, threshold characters were assumed to be influenced by an underlying continuous variation of the effects of multiple alleles (Dempster and Lerner 1950). This underlying distribution determines how liable it is that a threshold trait will be expressed. Thus, from a quantitative genetic perspective (sub-section 3.1.3), the expression of discontinuous phenotypes can be modelled as a threshold trait having an underlying normal distribution for liability (Falconer and Mackay 1996).

For many species, it seems that both genes and the environment play a role in determining whether individuals exceed or do not exceed thresholds for alternative phenotypes. For these situations, the evolutionary significance of thresholds has benefitted from conceptualizing threshold-type responses as sigmoidal reaction norms (section 3.4; Figure 8.6). Being part of a norm of reaction, the threshold is genetically determined, heritable, and responds to selection, but the probability that the threshold is exceeded, or triggered, depends on an environmental variable or trait-based cue.

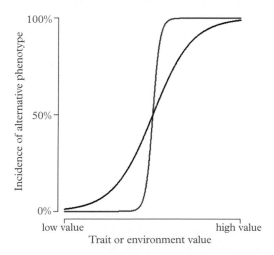

Figure 8.6 *Threshold norms of reaction describe the incidence of phenotypic expression as a sigmoidal function of a cue, i.e. an intrinsic individual trait or an extrinsic environmental factor. The threshold is practically defined at the 50 per cent incidence of expressing the alternative phenotype. A steeper slope (red line) reflects higher sensitivity to the cue.*

As noted in section 8.3, and discussed more generally in sections 3.3 and 3.4, it is unlikely that reaction norms will be genetically monomorphic, which would imply an absence of genetic variation in thresholds. Rather, different genotypes can be expected to have different reaction norms, simply expressed by a series of sigmoidal reaction norms. The pattern depicted in Figure 8.7, for example, would be indicative of stabilizing selection (i.e. selection that favours intermediate trait values by selecting against extreme trait values), a reasonable default expectation in the absence of additional information.

Selection might also be disruptive (favouring divergent or extreme trait values by selecting against intermediate trait values), such that each alternative phenotype is associated with a different threshold or set of thresholds (Figure 8.8). The key illustrative

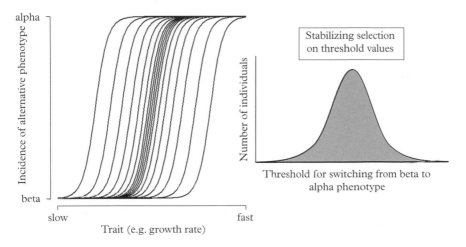

Figure 8.7 *Stabilizing selection for threshold reaction norms.*

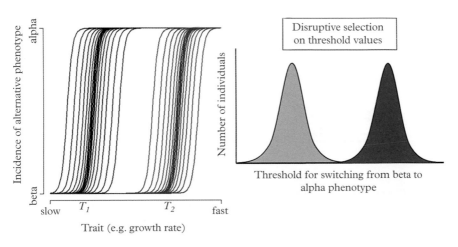

Figure 8.8 *Disruptive selection for two threshold reaction norms, centred at T_1 and T_2.*

point to be taken from Figures 8.7 and 8.8 is that threshold traits are very likely to be genetically variable within populations and subjected to different forms of selection, just as non-threshold traits are.

8.5.2   Thresholds for alternative strategies under selection

As discussed in section 8.4, different genetic architectures can set individuals on different long-term, developmentally constrained trajectories, leading to alternative strategic means of achieving successful reproduction. Being developmentally constrained, these alternative phenotypes typically have little or no probability of being reversed. Once a threshold that alters organismal development has been attained, individuals cannot flexibly change their developmental path and switch between alternative phenotypes.

The study of polyphenism (two or more distinct phenotypes are produced by the same genotype) has provided fertile ground for the study of genetic thresholds in insects (e.g. Walker 1986). Male European earwigs (*Forficula auricularia*), for example, produce forceps of dimorphic variability (Tomkins and Brown 2004). Those with long forceps actively use them for courtship and fighting and are more likely to guard females than males with short forceps, who must obtain fertilizations by sneaking. The greater the dietary nutrition of the developing nymph, the longer the pronotum (the dorsal exoskeletal plate of the earwig's first leg-bearing segment) and the longer the forceps (Figure 8.9(a)). The frequency of these alternative strategies can differ considerably among populations, the incidence of long-forceps males ranging from 8 to 45 per cent on Farne Islands off the eastern UK (Figure 8.9(b)). Common-garden experiments (section 3.4) have provided evidence that population differences in earwig thresholds are genetic (Tomkins and Brown 2004).

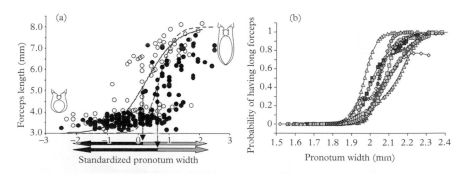

Figure 8.9 *Threshold reaction norms for island populations of European earwig. (a) The sigmoidal relationship between male size (pronotum width) and forceps length differs between Bass Rock (open circles) and Knoxes Reef (closed circles), leading to population differences in thresholds (vertical arrows) and proportions of each strategy (shaded sections of horizontal arrows). (b) Population differences in threshold reaction norms among 11 earwig populations on Farne Islands, UK.*

Figures from Tomkins and Brown (2004). Reprinted by permission from Springer Nature.

Dung beetles (family Scarabaeidae) have long provided excellent examples of alternative reproductive strategies (Arrow 1951; Emlen 1996). Males in several thousand species express dimorphism in the length of horns on their heads. Some produce relatively long horns while others produce rudimentary or no horns. Horned males are relatively large, fight one another for access to females, and defend access to female burrows. Smaller hornless males obtain fertilizations by sneaking, either stealing past horned males or avoiding them by constructing side tunnels that intersect with the female's burrow.

The expression of these alternative reproductive strategies is governed by a threshold norm of reaction between body size and horn length (Eberhard 1982). Emlen (1996) found that the reaction norm in *Onthophagus acuminatus* is heritable and responds to selection. After seven generations, the experimental populations selected for longer horns had their threshold shifted to smaller body sizes (increasing the probability that horns will be produced), whereas as those selected for shorter horns had their thresholds shifted to larger body sizes (Figure 8.10).

In addition to these earwig and beetle examples, there is considerable evidence in other insect groups of genetic variation in reaction-norm thresholds for alternative reproductive strategies, including wing dimorphism in sand field crickets (*Gryllus firmus*; Fairbairn and Yadlowski 1997) and fighter/scrambler strategies in the bulb mite *Rhizoglyphus echinopus* (Buzatto et al. 2012).

Among vertebrates, fishes provide instructive examples of threshold reaction norms. One of the most extreme sets of alternative strategies is evident within populations of Atlantic salmon (*Salmo salar*) (Jones 1959; Hutchings and Myers 1994; Vladić and Petersson 2015) (Figure 8.11). There are generally two types of males. Anadromous

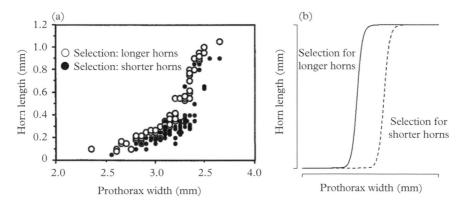

Figure 8.10 *In male horned beetles* (Onthophagus acuminatus), *alternative strategies are governed by a threshold reaction norm between horn length and body size, as measured by the prothorax (anterior-most segment, bearing the first pair of legs). (a) After selection, data for long- and short-horned males were such that (b) reaction norms for longer-horned males shifted to narrower prothoraxes and those for shorter-horned males to wider prothoraxes.*

Re-drawn from Emlen (1996). Reprinted by permission from John Wiley and Sons.

Figure 8.11 *Spawning Atlantic salmon. Anadromous female (left), anadromous male (right), and seven mature male parr, scaled to reflect the body sizes of parr relative to that of anadromous males in many Canadian populations.*

(seaward migrating) males attain large sizes because of the rapid growth they experience during one or more years at sea before returning to fresh water to spawn. By contrast, small male 'parr' mature in fresh water in the absence of a seaward migration; relatively few parr that actively participate in spawning subsequently migrate to sea and return to spawn (Hutchings and Myers 1994). Compared with the larger (>1000 g) and older (4–8 yr) anadromous males, mature male parr can reproduce at sizes orders of magnitude smaller (10–150 g) and at much younger ages (1–2 yr). Females do not mature as parr because of the massive fecundity cost—the production of <20 eggs as opposed to several thousand eggs—of reproducing at such a small size.

Prior to spawning, one or more anadromous males court a single anadromous female while she excavates a nest depression in the river substrate. During this period, mature parr position themselves close to the bottom substrate immediately behind the female's vent. Although parr will compete with one another for a position closest to the female's vent, they often swim little, so as not to attract the attention of the anadromous pair (the consequences of which can be fatal for parr). When the female extrudes eggs into the nest depression, parr dart in and release sperm in competition with anadromous males. As a group, based on laboratory and field studies, parr fertilization success can vary between nil and 100 per cent (Taggart et al. 2001; Piché et al. 2008). Multiple parr (more than ten in some cases) are capable of fertilizing the eggs of a single nest under natural conditions (Taggart et al. 2001; Weir et al. 2010). At the individual level, parr fertilization success tends to be low (~ five per cent of the eggs in a nest) and highly variable (Jones and Hutchings, 2001, 2002).

From a life-history perspective, the fitness consequences of maturing as either a parr or an anadromous male are clear, providing an excellent empirical example of the trade-offs of early versus delayed maturity discussed in section 4.1. For parr, by maturing at a young age and avoiding the high-mortality marine environment, the probability of surviving to reproduce is more than 100 times greater than that of anadromous males whose primary size-related fitness advantage is realized by higher, less variable fertilization success.

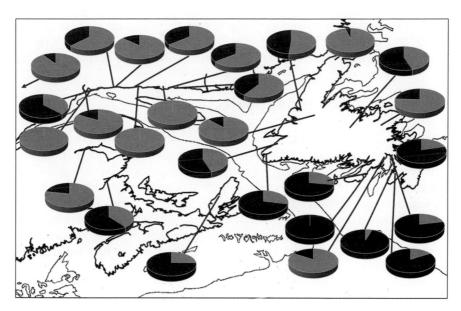

Figure 8.12 *Incidence (black) of the mature parr strategy among two-year-old males in 28 populations of Atlantic salmon in eastern Canada.*

Data from Myers et al. (1986).

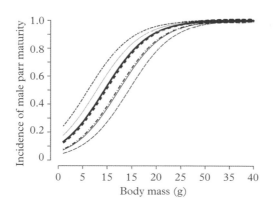

Figure 8.13 *Threshold norms of reaction between the proportional incidence of male parr maturity and individual growth (mass at seven months) in Atlantic salmon. Each reaction norm represents either a population or a cross between populations. From Piché et al. (2008).*

Reprinted by permission from the Royal Society.

The frequency of the mature male parr strategy differs considerably among populations (Figure 8.12). Adoption of the parr strategy depends on the attainment of a genetically determined threshold (Hutchings and Myers 1994; Piché et al. 2008). Those whose growth rate exceeds the threshold mature as parr; those whose growth rate does not exceed the threshold migrate to sea and mature as anadromous males (Figure 8.13).

It has been hypothesized that the parr and anadromous male strategies are governed by separate thresholds within populations in a manner consistent with Figure 8.8 (Hutchings and Myers 1994), a hypothesis supported by genomics research (Lepais et al. 2017).

8.6   Frequency-Dependent Selection

Alternative mating strategies are genetically based either through the existence of genetic polymorphisms or genetically different threshold norms of reaction. For simplicity (a simplicity with empirical support; section 8.4), strategies can be thought of as representing different alleles at the same locus. To persist, alternative strategies must have, on average, equal fitness (Shuster and Wade 2003) and the populations in which they are expressed cannot be invaded by individuals adopting other strategies (Maynard Smith 1982; Parker 1984). Thus, we would expect the alternative alleles—and the alternative strategies they underlie—to fluctuate around frequencies that are evolutionarily stable within populations.

Under these circumstances, the type of selection experienced by alternative strategies is presumed to be some form of negative frequency-dependent selection (Partridge 1988) (Figure 8.14). This type of selection occurs when the fitness of a phenotype decreases as it becomes increasingly common in a population. One reason for this could be increased competition among individuals adopting the same strategy. The higher the frequency of individuals adopting one strategy, the greater the competition amongst those individuals for access to mates, the lower the competition amongst individuals adopting the alternative strategy, and the higher the latter's fitness. The evolutionarily stable frequencies (ESF) of the two strategies occur at the point of intersection of the fitness functions at which individuals adopting each strategy have, on average, equal fitness (Figure 8.14).

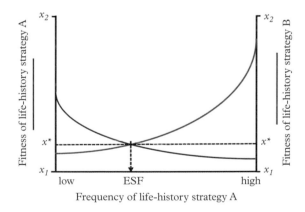

Figure 8.14 *Under negative frequency-dependent selection, the fitness of each strategy declines as the frequency of the strategy increases. The evolutionarily stable frequencies (ESF) of the two strategies occur at the point of intersection of the two fitness functions at which individuals adopting each strategy have, on average, equal fitness (x^*).*

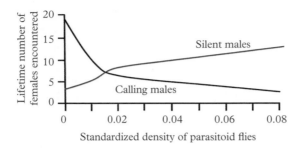

Figure 8.15 *Model-based fitness of calling and satellite (silent) male Texas field crickets as a function of increasing density of an acoustically orienting parasitoid fly.*

Based on data reported by Walker and Cade (2003).

Although depicted as a single value in the figure, in reality the ESF will likely fluctuate over time.

The model of negative frequency-dependent selection provides an empirically defensible explanatory framework for the maintenance of alternative strategies in species as taxonomically diverse as the marine isopod *P. sculpta* (Shuster and Wade 1991), side-blotched lizards (Sinervo and Lively 1996), and Atlantic salmon (Hutchings and Myers 1994).

Maintenance of alternative reproductive strategies need not always be governed by negative frequency-dependent selection within the same species. Inter-specific interactions can also play a significant role. In Texas field crickets (*Gryllus texensis*), males acquire mates either by acoustic signals (calls) or by being silent and then intercepting or searching for females (Cade and Cade 1992). The calling/silent dichotomy is heritable and responds to selection (Cade 1981).

Calling males are subject to attack by an acoustically orienting parasitoid fly, *Ormia ochracea*, after which reproductive impairment and death shortly follow for these males. Walker and Cade (2003) modelled how the fitness of calling and silent males varies with the density of parasitoid flies. As fly density increases, the fitness of calling males declines rapidly, whereas the number of females encountered by silent males increases (Figure 8.15). The model identifies a fly density at which the fitness of calling and silent males is equal, suggesting that parasitoid flies can be integral to the maintenance of alternative strategies in some crickets.

8.7   Clarity in Terminology

A dizzying array of names has been used to describe within-population differences in mating, reproductive, and parental-care activities among individuals. Examples include sneaker, cuckolder, parasite, bourgeois, scrambler, courter, satellite, caller, silent interloper, territorial, guarder, searcher, fighter, resource defender, usurper, spermatophore capper, female mimic, extra-pair copulater, clutch pirate, roamer, courser, coercer, surreptitious mater, egg dumper, harem guarder. Although these terms may have specific meaning for those who coined them, many refer to the same phenotype but for different species. Does extra-pair copulation in birds truly differ from sneak fertilizations in insects, lizards, and fishes?

Generalities have been sought by referring to alternative mating and reproductive phenotypes as either tactics or strategies. Two reviews have been highly influential.

In *Mating systems and strategies,* Shuster and Wade (2003) argue that for an alternative phenotype—a strategy—to successfully invade a population, there must be heritable variation for the traits underlying the phenotype. To persist in the population, both the novel and conventional strategies must be, on average, equally fit (otherwise the less fit phenotype would be selected against and eliminated). This must be true, even if the strategies occur at unequal frequencies. (This latter point is an important one. A persistent misconception is that alternative strategies need to co-occur at equal frequencies to have equal fitness.) In Shuster and Wade's (2003) conceptual and empirical framework, alternative strategies are genetically based, heritable, and directly subject to natural selection.

Taborsky et al. (2008), however, find the semantic borders between strategy and tactic to be vague and flexible. In *Alternative reproductive tactics,* they do not see utility in semantically distinguishing genetically based from non-genetically based phenotypes, remarking 'virtually all phenotypic traits are the product of genotypic *and* environmental influence [their *italics*]' (Taborsky et al. 2008: 3). They describe all discontinuous behavioural and other traits presumed to maximize fitness in the context of reproductive competition as alternative reproductive tactics.

Persuasive as Taborsky et al.'s (2008) arguments can be, it is empirically evident that not all phenotypic traits are heritable. And the conceptual disadvantages of a 'one-size-fits-all' approach—calling everything either a strategy or a tactic—do not outweigh the perceived terminological benefits. Inferring alternative phenotypes as targets of selection when they have no demonstrable heritable basis is problematic at best.

The application of tactics and strategies to describe alternative phenotypes remains unhelpfully inconsistent, contributing to a confusing, distracting, often self-contradictory use of terms. In 2020, I reviewed a multi-authored manuscript which concluded that age at maturity differs within populations because of 'the presence of alternative life history strategies such as alternative reproductive tactics'. A 2020 review of crustaceans (Garner and Neff 2020) considered a usurper (one who replaces a guarding male) to be a tactic that can take the 'forms' of sneaking and female mimicry. In contrast, Shuster and Wade (2003) identify usurpers, sneakers, and female mimics as alternative strategies.

There is incontestable logic in using words in a manner consistent with their historical and contemporary meaning. Borrowing from military and economic terminological analogs (the former dating at least to Sun Tzu's 2 500-year-old *The art of war*), strategies have long represented overall plans or frameworks to achieve long-term objectives (such as maximization of fitness). For semantic clarity, this book favours the perspective that alternative phenotypes be termed strategies if they are known to be a consequence of genetic polymorphisms or heritable threshold reaction norms. Tactics, on the other hand, represent flexible, reversible, short-term responses to prevailing conditions in support of a strategy.

8.8   Summing Up and a Look Ahead

Life-history differences are pervasive within populations. A subset of this variability is reflected by discontinuous variation in behavioural, developmental, and/or morphological traits during breeding. These mating and reproductive phenotypes—evident across taxonomically diverse species—are usually expressed as two alternatives (although some species express more than two). Most are expressed by males and are manifest by an alpha and a beta phenotype. Alpha males are typically larger, territorial, aggressive, and overt in attracting potential mates. Alpha males often obtain copulations by defending access to females and guarding resources necessary for reproduction. In contrast, beta males commonly obtain fertilizations by sneaking, adopting a satellite or searching behaviour, or mimicking females in colour and/or behaviour.

Alternative phenotypes can be distinguished as short-term, often reversible behavioural tactics or as longer-term, developmentally constrained, genetically influenced strategies. Their expression is highly dependent on the attainment of a threshold based on learning, experience, environmental quality, genetic architecture, or some combination thereof. Threshold reaction norms have proven valuable in providing theoretical and empirical constructs for understanding the evolutionary stability of alternative mating phenotypes. If there is one glaring deficiency in this research from a life-history perspective, it lies in the paucity of studies that have explored the consequences of alternative tactics and strategies to age-specific schedules of survival (l_x) and fecundity (b_x).

Thus far, the chapters of this book have provided a theoretical and empirical basis for understanding the evolution of life histories. Chapters 9 and 10 examine how knowledge of life histories and their evolution can be applied in matters pertaining to conservation and exploitation.

9

Applications: Conservation Biology

9.1 Three Paradigms

The field of conservation biology embraces three research frameworks or paradigms. The 'small-population' paradigm focuses on populations and species sufficiently low in number or density that their extinction probability is substantially increased. By virtue of their absolutely or relatively (say, to carrying capacity) small sizes, their fate is more likely determined by stochastic events or processes (see sub-section 6.3.4) than by life history. (On a terminological note, the words 'population' and 'species' tend to be used interchangeably in the literature on conservation biology; that practice will be followed in this chapter.)

The foundation of the second framework—the 'declining-population' paradigm—is a descriptive narrative of how species respond to human-induced disturbances and threats. It drives efforts such as the joint work by the World Wildlife Fund and the Zoological Society of London whose *Living Planet Index* (livingplanetindex.org) tracks global changes in the sizes of more than 20 000 vertebrate populations. This second paradigm explores reasons why declining populations might decline further, stabilize, or increase. Fundamental to this work are threat assessments of the vulnerability of species to human stressors, the most prominent being evaluations of extinction probability for which the IUCN (International Union for Conservation of Nature) is at the global forefront. As of June 2021, the IUCN had evaluated more than 134 400 species (iucnredlist.org). Species vulnerability assessments to extinction, over-exploitation, and climate change have provided fertile ground for predictions that account for life-history differences among species.

The 'recovering-population' paradigm of conservation biology examines how the rate and probability of population increase is influenced by factors other than threat mitigation, including life-history traits. Threat abatement alone need not always be sufficient to enable recovery.

As the chapter unfolds, it will become evident that life-history traits and per capita population growth (r) figure prominently in many species vulnerability and recovery assessments (e.g. Neubauer et al. 2013). Although links between life histories and r_{max} were established by the mid-1900s (Chapter 1), it was not until the late twentieth century

A Primer of Life Histories: Ecology, Evolution, and Application. Jeffrey A. Hutchings, Oxford University Press. © Jeffrey A. Hutchings 2021.
DOI: 10.1093/oso/9780198839873.003.0009

size over the longer of the most recent ten years or three generations (which for cod

that life-history traits were overtly considered or formally applied in evaluations of species extinction. Their application was inadvertently triggered by a particular group of species threat assessments made by the IUCN in the mid-1990s. This chapter begins with a telling of how this came about.

The chapter then explores the empirical basis for potential life-history correlates of species decline. Central to these efforts is a consideration of how life histories influence a population's resistance to external disturbance and its resilience following depletion. (Resistance refers to a population's ability to withstand, or resist, environmental change. Resilience reflects recovery potential.) Examples are provided of vulnerability assessments of species to extinction, over-exploitation, and climate change to illustrate how life-history traits have figured into these assessments.

The chapter closes with a focus on life-history correlates of recovery. Here, in addition to considering the theoretical basis for why life-history traits might be correlated with recovery potential, attention is directed to how variability in life histories within and among populations can affect the temporal stability of populations and species.

9.2   A Stimulus for Incorporating Life History to Assess Species Vulnerability

A primary stimulus for evaluating species vulnerability through a life-history lens stemmed from a controversy that arose concerning threat assessments of marine fishes. Following publication of a highly influential paper that proposed quantitative methods for assessing extinction risk (Mace and Lande 1991), the IUCN sought to rectify the low number of assessed marine fish species. Applying a set of criteria approved in 1994, the organization reported assessments of ~ 150 marine fish species in 1996 (Hudson and Mace 1996).

The fly in the ointment was that many of these fish species were commercially valuable. Disagreement was inevitable. It often is when conservation and commercial interests collide. The world's largest conservation organization was perceived to be challenging those whose purview was to assess or manage some of the world's largest fisheries. What was interesting was how central a role life history was to play in arguments against the listing of several species.

It is useful to recall the circumstances under which the IUCN's evaluations of marine fishes were initiated. The world had recently borne witness to several biologically, ecologically, and socio-economically devastating fishery collapses. Most prominent was that of Atlantic cod throughout the North Atlantic in the early 1990s (Myers et al. 1996). In May 1996, cod was one of several marine fish species to be globally assessed by the IUCN as facing heightened risk of extinction (Hudson and Mace 1996). Cod was evaluated as Vulnerable, based on a >20 per cent reduction in estimated population size over the longer of the most recent ten years or three generations (which for cod populations would have been about 20–30 years). Other marine fish species were assessed as Endangered (>50 per cent decline) and others still as Critically Endangered (>80 per cent decline).

Reaction from many fisheries scientists and fisheries managers was swift and not complimentary. Arguments were put forth to explain why marine fishes should not be subjected to the same criteria that the IUCN uses to assess terrestrial species such as birds and mammals (for discussions, see Musick 1999; Hutchings 2000, 2001; Reynolds et al. 2005).

Firstly, it was argued that substantial population declines need not always be of conservation concern. Rather, for commercially harvested fish species, they can be the result of a management strategy intended to reduce population size to a level at which the maximum sustainable yield (MSY) might be obtained (see Chapter 10). As noted in Chapter 1 (Figure 1.5), under the simplest set of assumptions, population growth rate ($\partial N/\partial t$) is maximized at 50 per cent of a population's carrying capacity. The greater the value of $\partial N/\partial t$, the greater a fishery's sustainable yield or catch. Put another way, if an unfished population is at or near carrying capacity, fishery managers might wish to reduce N by at least 50 per cent. But this would be a magnitude of decline that, if realized rapidly, might be sufficient to produce a threat assessment of Vulnerable or Endangered by the IUCN.

Strictly speaking, the fishery critics had a point. A counter-point was that few if any of the species assessed by the IUCN in 1996 had been in an unfished state when the population size declines had been initiated. Given a lack of knowledge as to what the population sizes were in the unfished states, there was no means of ascertaining how many, if any, of the species assessed by the IUCN had been deliberately fished under a management strategy with the purpose of achieving MSY.

A second argument was that marine fishes should not be assessed by the same criteria that the IUCN applied to other vertebrates because the life histories of fishes differed so much from their terrestrial counterparts. These life-history differences were perceived to confer high-fecundity fishes much greater recovery potential, or resilience, than low-fecundity species on land (Musick 1999; Powles et al. 2000; DeMaster et al. 2004). Many marine fishes (other than chondrichthyan sharks, skates, and rays) are capable of producing hundreds of thousands if not millions of eggs per breeding season, far in excess of the numbers of offspring per clutch produced by other vertebrates (Figures 2.4 and 7.1). Fisheries scientists advising the UN's Food and Agriculture Organization opined that greater fecundity would tend to make aquatic species more resilient to depletion and result in a lower risk of extinction (FAO 2000).

One implication of this supposition that higher fecundity confers lower extinction probability and greater recovery potential is that r_{max} is higher in marine fishes than in terrestrial vertebrates. However, as large-scale analyses have found, r_{max} does not differ, on average, between marine teleost (bony) fishes and terrestrial mammals (Figure 9.1). r_{max} is significantly lower in chondrichthyan fishes and significantly lower still in marine mammals.

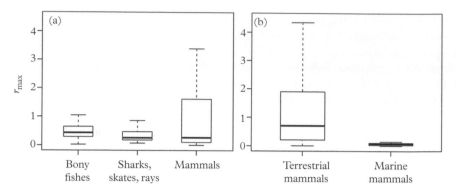

Figure 9.1 *Median maximum per capita population growth rates (r_{max}) for (a) bony fishes (teleosts), sharks, skates, and rays (chondrichthyans), and mammals (terrestrial and aquatic species combined) and for (b) terrestrial mammals and marine mammals. The central line of the boxplots indicates the median, boxes span the interquartile range, and whiskers encompass values less than 1.5 times the interquartile range away from the box. Median r_{max} values and confidence intervals (CI) for each group (n is number of species): bony fishes = 0.432 (CI 0.354 – 0.531; n = 47); sharks, skates, rays = 0.256 (CI 0.207 – 0.298; n = 82); mammals = 0.266 (CI 0.138 – 0.706; n = 70); terrestrial mammals = 0.706 (CI 0.270 – 1.286; n = 54); marine mammals = 0.073 (CI 0.030 – 0.095; n = 16).*

From Hutchings et al. (2012).

9.3   Life-History Correlates of r_{max}

One priority of conservation biology is to assess the vulnerability of species to human-induced environmental change. Vulnerability can be defined in relation to a species' ability to withstand negative effects produced by the external environment. A species that is highly vulnerable is one that has low resistance and is highly susceptible to disturbance. Considerable effort has been expended in evaluating species' vulnerability to external threats. These assessments are conducted for a variety of reasons such as evaluating vulnerability to extinction, over-exploitation, climate change, and human-driven habitat alteration.

A consideration of life history was evident in the IUCN's 1994 incarnation of their criteria used to assess conservation status. Although life-history traits per se are not explicitly mentioned, one criterion uses rate of population decline to determine status (this is the same criterion mentioned in section 9.2). The time over which the decline is measured is the most recent ten years or the most recent three generations for which data are available, whichever time frame is longer.

Scaling the rate of decline by generation length accommodates species differences in their rate of turnover. This makes sense from a life-history perspective. Rate of turnover is directly related to r_{max}. The higher the average genotypic value of r_{max} in a population, the faster the rate at which the average genotype passes its genes to future generations and, thus, the higher the rate of population turnover (section 4.1).

Many assessments identify 'productivity' as a measure against which vulnerability can be assessed. Definitions of productivity differ considerably, a problem recognized since at least the 1930s (Macfadyen 1948). The vulnerability assessments considered in this chapter treat productivity as being synonymous with maximum per capita rate of population growth, or the capacity of a population to grow or increase in abundance, particularly following depletion. A key inference is that the higher the productivity of a species, the lower its vulnerability to external threats.

As indicated by Figure 9.2, not only can species differ considerably in r_{max}, but so can populations within a single species. However, while justification for using r_{max} is theoretically strong, it can be difficult to apply in practice for the simple reason that r_{max} is challenging to estimate for natural populations. This has led to a search for more readily measurable correlates of r_{max}.

Let's return to the fact that the IUCN bases its population-decline criterion on generation time. Recall that generation time is the average age of parents of a single cohort or year class (e.g. all of the young born in 2001) (section 5.4). The older the age at maturity, the longer the generation time (Chapter 5). In other words, the IUCN's population-decline criterion implicitly accounts for species differences in age at maturity. But is age at maturity—or for that matter, any other life-history trait—a reliable metric of r_{max}? A cursory glance at Figure 9.2 suggests that it might be. Among the species considered, as age at maturity (α) increases, r_{max} declines.

An analysis of a large dataset (199 species) of terrestrial and aquatic vertebrates provided some answers. The vertebrates examined included 47 species of bony (teleost) fishes, 82 species of chondrichthyan fishes (sharks, skates, rays), and 70 species of

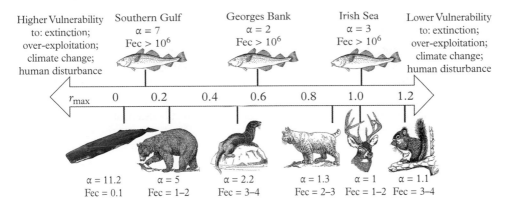

Figure 9.2 *Many species and population assessments implicitly or explicitly assume that higher r_{max} is associated with reduced vulnerability to extinction, over-exploitation, climate change, and other human disturbances. Here are estimates of r_{max}, age at maturity (α in yr), and fecundity (Fec = average number of offspring per year) for a selection of vertebrates. Above arrow: Atlantic cod* (Gadus morhua) *populations from Southern Gulf of St. Lawrence (Canada), Georges Bank (Canada/US), and Irish Sea. Below arrow (l − r): sperm whale* (Physeter macrocephalus), *black bear* (Ursus americanus), *river otter* (Lontra canadensis), *bobcat* (Lynx rufus), *white-tailed deer* (Odocoileus virginianus), *red squirrel* (Tamiasciurus hudsonicus). *Estimates of r_{max} are from Myers et al. (1997) and Hutchings et al. (2012).*

Line drawings are in the public domain except the red squirrel © Can Stock Photo/Birchside.

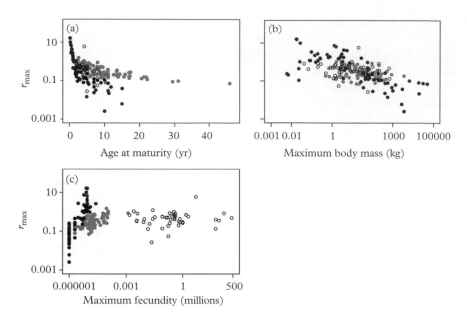

Figure 9.3 *Estimates of* r_{max} *for mammals (black), bony fishes (open), and sharks, skates and rays (grey). There are negative relationships (p <0.05) between* r_{max} *and (a) age at maturity and (b) maximum body size. These negative correlations also exist for each group when considered separately, except for bony fishes in (b). There is no association between* r_{max} *and (c) fecundity within either group of fishes. Mammals that produce a single offspring per breeding period (black circles at the left end of the x-axis) have lower* r_{max} *than vertebrates that produce more than one individual per breeding period.*

From Hutchings et al. (2012).

mammals, including 16 aquatic species (whales and seals). One objective of this work was to quantify potential correlations between r_{max} and each of three traits for which data on at least one are often readily available: age at maturity, maximum body size, and fecundity. The analyses indicated that age at maturity is a reliable metric of r_{max}, particularly when age at maturity is less than 10 yr (Figure 9.3(a)). The same set of data was examined to explore potential links between body size and r_{max}. Again, the results suggested a very strong link (Figure 9.3(b)). As body size increases, r_{max} decreases, confirming a conclusion dating to the mid 1970s (Fenchel 1974).

The dataset was also analysed to examine whether r_{max} increases with increasing fecundity. This provided an opportunity to empirically test the presumption that higher fecundity in fishes is associated with lower vulnerability to external threats (section 9.2). Contrary to this presumption, there is no association between r_{max} and fecundity in fishes, although there appears to be a positive link in mammals because of the low r_{max} associated with the production a single offspring per breeding period (Figure 9.3(c)). The analysis provides confirmation of Cole's (1954) modelling output regarding fecundity, and arguments made subsequently (e.g. Sadovy 2001; Dulvy et al. 2003), that high fecundity has no discernable effect on r_{max} or, by extension, species vulnerability.

9.4   Species Vulnerability Assessments

9.4.1   Extinction

It has been argued that the IUCN takes an unduly minimalist approach in accounting for life history in its threat assessments (Musick 1999; DeMaster et al. 2004). That said, if you wish to account for a single life-history trait, then age at maturity, as reflected by generation time in the IUCN's population-decline criterion, is perhaps the most defensible, being the trait to which r_{max} tends to be most sensitive (Lewontin 1965). The IUCN's argument against taking a more detailed life-history approach lies in the organization's wish to achieve consistency and generality in the application of its assessment criteria across all species (Collen et al. 2016).

But if the goal is to assess vulnerability within a specific taxonomic group, particularly a specific population, there are advantages in applying a more detailed approach. One justification for doing so is that generation time (or age at maturity), on its own, might not always provide a sufficient metric of species differences in resistance and resilience to disturbance (Lake 2013). As a reminder, resistance refers to the ability of a population to withstand, or resist, a change in the environment with potential to affect population viability. In a life-history context, these are changes arising from environmental stressors that act on age-specific rates of survival and fecundity, meaning that they have potential to affect $r_{realized}$ and r_{max}. Resilience, being the capacity to recover from population declines imparted by a threat or disturbance, is also related to $r_{realized}$ and r_{max}. (Somewhat confusingly, resistance is sometimes considered a component of resilience, e.g. Capdevila et al. 2020.)

Definitions aside, resistance and resilience are often thought to reflect the productivity of a species which inevitably leads, once again, to a consideration of r_{max}. Since Cole's (1954) initial work on linking life-history traits with a population's intrinsic rate of increase, considerable effort has been undertaken to identify life-history correlates of r in a conservation context.

The most widely applied sets of life-history correlates of vulnerability may be those developed for fishes. From a conservation perspective, marine fishes garner considerable attention. One reason for this is that fishes comprise the largest group of wild, undomesticated organisms that humans regularly consume as food. A second reason pertains to the observation that the primary, often only, threat to marine fishes is overexploitation. Unlike freshwater fishes and terrestrial plants and animals, marine fishes are not generally considered to be threatened by habitat alteration or species invasions (coral reef fishes being an exception). A third advantage to assigning quantitative categories of vulnerability to marine fishes is that, because of their commercial importance, there tends to be a very considerable amount of life-history data available across a phylogenetically broad swathe of species.

Early attempts to assess marine fish vulnerability represented a response by fishery scientists to the IUCN's 1996 threat assessments (section 9.2). The American Fisheries Society (AFS) adopted a classification system based on r_{max} that distinguishes four levels of productivity (Musick 1999), ranging from very low (r_{max} <0.05) to high (r_{max} >0.50)

Table 9.1 *Metrics of productivity as applied to marine fishes by the American Fisheries Society (AFS). The assigned level of productivity determines the decline threshold (70 per cent, 85 per cent, 95 per cent, 99 per cent) that must be met, over the longer of three generations or ten years, before a species or population is assessed as vulnerable by the AFS.*

| Trait | Productivity | | | |
	Very low	Low	Medium	High
r_{max}	<0.05	0.05–0.15	0.16–0.50	>0.50
Age at maturity	>10 yr	5–10 yr	2–4 yr	<1 yr
von Bertalanffy k	<0.05	0.05–0.15	0.16–0.30	>0.30
Maximum age	>30 yr	11–30 yr	4–10 yr	1–3 yr
Fecundity	<10	10–100	101–10 000	>10 000
Decline threshold to trigger an assessment of *vulnerable*	70 per cent	85 per cent	95 per cent	99 per cent

(Table 9.1). Three life-history traits are included: fecundity, age at maturity, and maximum age. The von Bertalanffy k coefficient (sub-section 2.3.2), a metric related to individual growth, is also included.

Application of the scheme begins with r_{max}. If estimates of this parameter are available, species are directly categorized as having very low, low, medium, or high productivity, based on the categories in Table 9.1. If estimates of r_{max} are not available, species productivity is determined by the four traits, as estimated for the species or population in question, ideally in an unexploited state.

Musick (1999) considered age at maturity (α) to be the most important correlate of r_{max}. Given that k tends to be negatively corrected with α, and maximum age positively correlated with α, these two traits were considered important metrics of productivity as well. Musick (1999) cautioned that low fecundity might be useful in flagging poor-productivity species but that high fecundity might give a misleading impression of productivity. In the absence of estimates of r_{max}, the productivity assigned to a species or population corresponds to the lowest productivity category for which data are available in the table. Once the productivity of a species has been assigned, specific decline thresholds must be exceeded (bottom row of Table 9.1) before a species or populations is considered at heightened probability of extinction and assessed by the AFS as *vulnerable*.

Leaving aside the defensibility of the numeric values of r_{max} and life-history metrics that delineate the AFS's decline thresholds, we can ask whether the traits included in this extinction assessment framework for marine fishes have merit. Given what we know about age at maturity and r_{max} (Figure 9.3(a)), we can conclude that α merits inclusion. Use of k can be defended because of its positive association with M (sub-section 2.3.2),

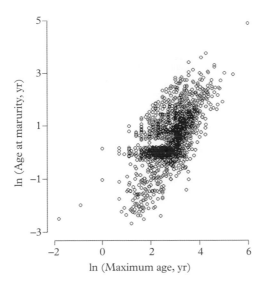

Figure 9.4 *Age at maturity is positively associated with maximum age in vertebrates: birds (n = 647 species); mammals (n = 743); reptiles (n = 63); amphibians (n = 69); teleost fishes (n = 254); chondrichthyan fishes (n = 106).*

Data source: de Magalhães and Costa (2009).

meaning that populations comprised of relatively fast-growing individuals experience correspondingly higher levels of natural mortality. There is also merit in including maximum age, given its positive association with age at maturity (Figure 9.4). The weakest link in the framework is the inclusion of fecundity.

The use of life-history traits to guide extinction vulnerability assessments has continued with efforts to develop quantitative criteria applicable to all terrestrial and aquatic species. A primary motivation was a perceived need to strengthen the quantitative rigour of science advice associated with the listing of threatened and endangered species in accordance with national legislation, such as the *Endangered Species Act* in the US and the *Species at Risk Act* in Canada (DeMaster et al. 2004; DFO 2007; Regan et al. 2009; Waples et al. 2013).

9.4.2 Exploitation

The field of fisheries conservation biology emerged during the decade of globally prominent population collapses in the 1990s. Daniel Pauly coined the phrase in his obituary of one of the founders of the field, Ransom Myers, defining the discipline as one 'devoted to identifying exploited fish populations and species threatened with extinction, and suggesting measures for rebuilding them, along with the ecosystems in which they are embedded' (Pauly 2007: 160).

Table 9.2 *Life-history metrics of species/population productivity used to assess the vulnerability to over-exploitation of marine fishes in 162 US fisheries (Patrick et al. 2010).*

Metric	Productivity		
	Low	Moderate	High
r_{max}	<0.16	0.16–0.50	>0.50
Age at maturity	>4 yr	2–4 yr	<2 yr
von Bertalanffy k	<0.15	0.15–0.25	>0.25
Maximum age	>30 yr	10–30 yr	<10 yr
Fecundity	<100	100–10 000	>10 000
Maximum size	>150 cm	60–150 cm	<60 cm
Natural mortality, M	<0.20	0.20–0.40	>0.40

Increased awareness of the negative consequences of over-fishing led to policies and regulations that espoused ecosystem-based fisheries management. Some countries enacted legislation in support of such an objective. The Australian *Environment Protection and Biodiversity Conservation Act 1999*, for example, requires managers to demonstrate that fisheries are ecologically sustainable (Hobday et al. 2011). These initiatives created a need for frameworks to assess the vulnerability of species directly targeted, and those incidentally caught, by exploitation. And if the assessments were to have meaningful relevance to the protection of vulnerable species and ecosystems, they needed to be applicable to data-poor species.

As with species-extinction evaluations, life-history traits proved invaluable as a means of assessing vulnerability to fishing. Patrick et al. (2010) developed one such vulnerability assessment, applying it to 162 fished populations. As with Musick (1999), they identified productivity (which they defined as the capacity to recover rapidly following depletion; another term would be resilience) as a key determinant of vulnerability to overfishing. Primacy was again given to r_{max} as a measure of productivity (Table 9.2).

Following Musick's (1999) lead, Patrick et al. (2010) included age at maturity (α), the von Bertalanffy growth coefficient k, maximum age, and fecundity. Their vulnerability assessment included two additional variables related to life history. One was maximum size which is positively correlated with maximum age (Figure 2.9). The other was natural mortality, i.e. death caused by factors other than fishing. Natural mortality (commonly designated as M) is an annual rate typically applied to the adult stage of life (age at maturity until death) and usually assumed to be constant from one age to the next. M directly reflects productivity insofar as populations that experience high rates of natural mortality require high levels of production to maintain abundance (Patrick et al. 2010).

The passing of Australia's *Environment Protection and Biodiversity Conservation Act* stimulated research to strengthen the application of ecosystem-based fishery management. One of these initiatives was an ecological risk assessment framework termed

Table **9.3** *Life-history attributes of productivity used to assess the susceptibility of marine fishes to exploitation as part of an ecological risk-assessment framework for the effects of fishing (Hobday et al. 2011).*

Attribute	Productivity		
	Low	Moderate	High
Age at maturity	>15 yr	5–15 yr	<5 yr
Maximum age	>25 yr	10–25 yr	<10 yr
Fecundity	<100	100–20 000	>20 000
Maximum size	>300 cm	100–300 cm	<100 cm
Size at maturity	>200 cm	40–200 cm	<40 cm

Ecological Risk Assessment for the Effects of Fishing (ERAEF) (Hobday et al. 2011). The ERAEF framework comprises a three-tiered approach. The initial tier is a qualitative assessment, based on expert judgement, of the impact of fishing on targeted species and those caught incidentally as bycatch. The impact on each species is scored, ranging from 1 (negligible) to 6 (extreme). Those scored 3 or higher move on to tier 2, where the productivity of each species (productivity being related to r; Table 9.3) and its susceptibility to capture by fishing are evaluated. The tier 2 assessment, termed a Productivity Susceptibility Assessment (Hobday et al. 2011), has since been applied to more than 1 000 fish populations (Hordyk and Carruthers 2018). Species ranked as medium or high risk in tier 2 are then subjected to a conventional fishery stock assessment to quantitatively specify risk.

It is the second tier, where species productivity is assessed, that life-history attributes enter the picture (Table 9.3). As with the vulnerability assessments of Musick (1999) and Patrick et al. (2010), the ERAEF incorporates age at maturity, maximum age, and fecundity. It also includes two measures of body size: the maximum length and the length at maturity. In terms of productivity thresholds, those for fecundity are similar to those applied by Musick (1999) and Patrick et al. (2010), whereas those for age at maturity differ considerably.

9.4.3   Climate change

If extinction assessments can be dated from the 1960s (the IUCN was founded in 1964) and fishing vulnerability assessments from the 1990s (e.g. Jennings et al. 1998), the twenty-first century has borne witness to vulnerability assessments of species to climate change.

The approach of combining information on sensitivity, exposure, and capacity to adapt to climate has informed climate-change vulnerability assessments for many groups of organisms, including plants (Keith et al. 2008; Matthews et al. 2011), freshwater fishes (Moyle et al. 2013), birds, amphibians, and corals (Foden et al. 2013), marine fishes and invertebrates (Hare et al. 2016), and sharks and rays (Chin et al. 2010).

There is broad agreement among conservation biologists of the need for life history to be factored into climate-change vulnerability assessments, given the influence that life histories have in influencing species productivity, resistance, and resilience (Glick et al. 2011; Foden et al. 2019). Pearson et al. (2014), for example, undertook a quantitative analysis of attributes predicted to cause species to be at high risk of extinction specifically due to climate change. The predictors included variables related to life history, habitat, dispersal, niche breadth, population structure, and recent trends in population size, occupied area, connectivity, and habitat use. Three predictors were specifically related to life history. These were r_{max}, generation time, and variability in age- or stage-specific survival (l_x) and fecundity (b_x) (this variation measures the tendency of a population to fluctuate in response to stochastic environmental change). Of the 21 predictors, generation time (ranked fourth) and r_{max} (ranked seventh) were among the most important determinants of climate-change vulnerability (the area occupied by a species was the highest predictor of extinction under climate change; the smaller the area, the greater the extinction probability).

The life-history attributes considered in species climate-change vulnerability assessments do not differ appreciably from those applied to marine fish assessments of extinction probability (Table 9.1) and fishing (Tables 9.2 and 9.3). Consider those in Foden et al.'s (2013) trait-based vulnerability assessments of the world's birds, amphibians, and corals ($n = 16\,857$ species; Table 9.4). Metrics of fecundity in marine fishes are evident in birds (clutch size) and amphibians. Maximum age and age at maturity in fishes are related to longevity (corals) and generation time (birds), respectively. Inclusion of growth rate in corals can be considered analogous to application of the von Bertalanffy growth coefficient (k) in marine fishes.

Table 9.4 *Correlates of high and low/lower vulnerability to climate change for birds, amphibians, and corals (compiled from data provided by Foden et al. 2013).*

	Vulnerability to climate change	
	High	Low/lower
Birds		
Clutch size (number of eggs)	≤2	>2
Generation time (years)	≥6	<6
Amphibians		
Annual number of eggs	≤50	>50
Corals		
Longevity (years)	≥50	<50
Maximum growth rate (mm yr^{-1})	≤30	>30

Table 9.5 *Life-history metrics of per capita population growth rate (r_{max}) used in a US National Marine Fisheries Service vulnerability assessment of marine fishes and invertebrates to climate change (Hare et al. 2016).*

	Sensitivity to climate change			
	Very High	High	Moderate	Low
r_{max}	<0.05	0.05–0.15	0.16–0.50	>0.50
von Bertalanffy k	<0.10	0.11–0.15	0.16–0.25	>0.25
Age at maturity	>5 yr	4–5 yr	2–3 yr	<2 yr
Maximum age	>25 yr	15–25 yr	11–15 yr	<10 yr
Natural mortality, M	<0.20	0.21–0.30	0.31–0.50	>0.50

In response to information needs required to fulfil national climate-change policy objectives, the US National Marine Fisheries Service initiated a process that culminated in a set of climate-change vulnerability assessments for 82 fish and invertebrate species inhabiting waters off the northeast coast of the US. The primary goal was to produce a practical and efficient tool for assessing vulnerability, based on factors likely to affect the exposure of species to climate change and the sensitivity of species to that exposure.

The initiative identified 12 attributes predicted to affect species sensitivity to climate change (Hare et al. 2016). One pertained to r_{max}. To assess species for which estimates of r_{max} were not available, life-history metrics of r_{max} were used to assign species to different levels of climate-change sensitivity (Table 9.5). Of the 12 biological sensitivity attributes that were applied, Hare et al. (2016) found that their species assessments were most sensitive to r_{max}.

9.5   A Comparison of Vulnerability Assessments

By the early twenty-first century, the evaluation frameworks that had been developed could be used to inform future initiatives from a life-history perspective. Albeit a numerically small sample, the assessment frameworks presented in this chapter are representative of many such initiatives, evidenced in part by commonalities in the life-history attributes used to assess vulnerability to climate change by birds, amphibians, and corals (Table 9.4; Foden et al. 2013) and those used to assess vulnerability to extinction, exploitation, and climate change in marine fishes (Table 9.6).

The information in Table 9.6 invites a comparative evaluation. Despite differences among frameworks, there are more commonalities than dissimilarities. Perhaps the most striking differences are in the productivity thresholds for some attributes. Those applied for age at maturity and maximum age in assessments of the ecosystem effects of fishing (EREAF; Hobday et al. 2011) differ considerably from the thresholds applied in other assessments in Table 9.6. These differences in thresholds draw attention to two key

Table 9.6 *Life-history attributes common to at least two of four evaluation frameworks to assess the vulnerability of marine fishes to extinction (Musick 1999), climate change (Hare et al. 2016), and fishing (1 = Patrick et al. 2010; 2 = Hobday et al. 2011). Abbreviations: agemat = age at maturity; vonB k = von Bertalanffy growth coefficient; maxage = maximum age; maxsize = maximum length; natmort = natural mortality (M).*

| Trait | Scheme | Vulnerability/sensitivity/susceptibility | | | |
		Very high	High	Medium	Low
r_{max}	Extinction	<0.05	0.05–0.15	0.16–0.50	>0.50
	Climate change	<0.05	0.05–0.15	0.16–0.50	>0.50
	Fishing 1		<0.16	0.16–0.50	>0.50
	Fishing 2		related to r	related to r	related to r
agemat	Extinction	>10 yr	5–10 yr	2–4 yr	<1 yr
	Climate change	>5 yr	4–5 yr	2–3 yr	<2 yr
	Fishing 1		>4 yr	2–4 yr	<2 yr
	Fishing 2		>15 yr	5–15 yr	<5 yr
maxage	Extinction	>30 yr	11–30 yr	4–10 yr	1–3 yr
	Climate change	>25 yr	15–25 yr	11–15 yr	<10 yr
	Fishing 1		>30 yr	10–30 yr	<10 yr
	Fishing 2		>25 yr	10–25 yr	<10 yr
vonB k	Extinction	<0.05	0.05–0.15	0.16–0.30	>0.30
	Climate change	<0.10	0.11–0.15	0.16–0.25	>0.25
	Fishing 1		<0.15	0.15–0.25	>0.25
	Fishing 2				
fecundity	Extinction	<10	10–100	101–10 000	>10 000
	Climate change				
	Fishing 1		<100	100–10 000	>10 000
	Fishing 2		<100	100–20 000	>20 000
maxsize	Extinction				
	Climate change				
	Fishing 1		>150 cm	60–150 cm	<60 cm
	Fishing 2		>300 cm	100–300 cm	<100 cm
natmort	Extinction				
	Climate change	<0.20	0.21–0.30	0.31–0.50	>0.50
	Fishing 1		<0.20	0.20–0.40	>0.40
	Fishing 2				

points. Firstly, given that life histories can differ among geographical regions (e.g. tropical versus temperate areas), the productivity/vulnerability thresholds can be expected to differ as well. Secondly, it is important to remember that the threshold determinations are often subjective and arbitrary, meaning that they are the product of expert opinion rather than rigorous quantitative analysis. Thus, the thresholds can differ when the experts differ.

In terms of commonality, all four assessment frameworks implicitly (Hobday et al. 2011) or explicitly use r_{max} as the overarching measure of vulnerability. The thresholds used to identify low, medium, and high r_{max} are largely indistinguishable, reflecting a degree of comfort by Patrick et al. (2010) and Hare et al. (2016) with Musick's (1999) classification. A second common feature is that each incorporates age at maturity and maximum age (or lifespan). These choices are highly defensible, given that r_{max} is tightly linked with age at maturity (Figure 9.3(a)) and maximum age (via correlation with age at maturity; Figure 9.4).

Three of the frameworks use the von Bertalanffy growth coefficient (k) and fecundity. Inclusion of the former can be defended, given its positive association with natural mortality (M) (Thorson et al. 2017). Following Hare et al. (2016), fecundity should be excluded from vulnerability assessments, including Productivity Susceptibility Assessments (Hordyk and Carruthers 2018), because of its absent or equivocal association with r_{max} (Figure 9.3(c)), and because taxonomic affiliation alone is sufficient to separate those fishes with very low fecundity (Chondrichthyes) from high-fecundity species.

Only two assessment frameworks (Patrick et al. 2010; Hare et al. 2016) included M and maximum size. Although M is widely acknowledged to be relevant to the assessment of species vulnerability, its absence in some frameworks can likely be explained by the difficulty in directly estimating natural mortality. The less-frequent use of maximum size is perhaps surprising, given its association with r_{max} (Figure 9.3(b); Fenchel 1974) and that large body size (along with late age at maturity) has been identified as a reliable predictor of vulnerability to fishing, based on a review of 15 studies by Reynolds et al. (2005).

9.6   Species and Population Recovery

9.6.1   Generalized approaches are uncommon

As section 9.4 illustrated, quantitative criteria relating to life history are routinely applied to assessments of large numbers of species to assess their vulnerability to human-generated threats. Such a generalized approach can infuse clarity and consistency to setting conservation priorities for taxonomically broad sets of species. But while a generalist approach has been favoured when addressing factors causing population decline and collapse, approaches to species recovery are much more likely to be very specifically focused on a single population or species. Further confounding a generalist approach is the fact that definitions of what constitutes recovery differ considerably, having been used to encompass a wide range of scenarios, from simply achieving the minimum conditions for species persistence, to full recovery of a species' ecological and evolutionary functionality within its ecosystem (Redford et al. 2011; Hutchings et al. 2012).

There is a perception that the only thing affecting the recovery of a species is the threat responsible for its decline. Remove or mitigate the threat and recovery will follow. While threat mitigation is clearly necessary to halt the decline of a species, threat abatement in and of itself is not always sufficient to result in recovery. Examples of species for which threat removal has not resulted in recovery include the orange-bellied parrot (*Neophema chrysogaster*) in southern Australia (Martin et al. 2012), the black-footed ferret (*Mustela nigripes*) in the US (USFWS 2016), and numerous marine fishes worldwide (Hutchings 2015).

9.6.2 Life-history correlates of recovery potential

Among the few classification schemes that assess recovery potential is one that simply defines recovery as a reversal of declines and achievement of predefined targets (these targets typically relate to metrics of persistence, such as abundance, distribution, and genetic/phenotypic variability). The framework includes life-history traits, changes in $r_{realized}$ and its variance, and a consideration of life-history variability within and among populations of the same species (Table 9.7). Although lacking the numerical thresholds evident in vulnerability assessments, this scheme identifies life-history attributes likely to

Table 9.7 *Correlates of impaired recovery (modified from Hutchings et al. (2012).*

Correlate	Impediments of recovery	Postulated influence on recovery
Life-history traits	Advanced age at maturity	Trait combinations associated with lower r_{max} will retard recovery
	Large size at maturity	
	Long generation time	
	Slow individual growth rate (ectotherms)	
	Low fecundity (excluding teleost fishes)	
$r_{realized}$	Declines with declining population size (Allee effect)	Populations that fall below the threshold at which Allee effects are expressed will have slower, more uncertain recovery
Variance in $r_{realized}$	Increased variance with declining population size	The greater the variance in $r_{realized}$ in small populations, the greater the uncertainty in the trajectory of recovery
Life-history variation among populations	Reduced variability	Reduced population variation can negatively affect recovery at the meta-population or species level (reduced portfolio effect)

impede recovery: advanced age at maturity, large size at maturity, long generation time, slow individual growth rate (ectotherms), low fecundity (excepting fishes).

Interestingly, one recovery attribute missing from Table 9.7 is natural mortality, M. A comparison of 55 marine fish populations that had and had not recovered revealed that life- history traits, when considered singly, were not correlated with recovery (Hutchings and Kuparinen 2017). However, when traits were considered in combination, to estimate natural mortality, recovered populations experienced higher M than populations that had not recovered. Populations comprising individuals adapted to high M tend to also have earlier maturity and attain smaller maximum sizes both of which are associated with high r_{max} (Figures 9.3(a) and (b)). However, although recovery potential appears to be greater for populations adapted to high M, short-term (one to three generations) increases in M within a population are likely to slow recovery (Swain 2011).

The recovery correlates in Table 9.7 go further than most vulnerability assessments by drawing explicit attention to how $r_{realized}$ changes with population size. Integral to the dynamics of small populations are Allee effects. These are manifest in declining populations when a threshold is reached beyond which $r_{realized}$ begins to decline, rather than continuing to increase, with further declines in population size (Figure 9.5(a)). The greater the magnitude of decline, the greater the likelihood that Allee effects will be manifest. Indeed, there is evidence for marine fishes and invertebrates that the greater magnitude of population reduction, the longer and more uncertain the recovery period (Hutchings 2000; Neubauer et al. 2013).

Allee effects offer a theoretical and empirical basis for identifying population-size thresholds below which recovery is increasingly likely to be impaired and uncertain (Courchamp et al. 2008; Hutchings 2015). However, making empirical determinations of where these population-size thresholds are located along the x-axis remains a challenge.

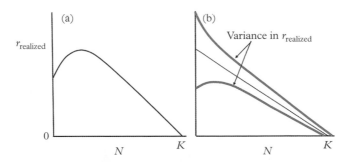

Figure 9.5 *Potential changes in $r_{realized}$ and its variance with changes in population size, N. (a) An Allee effect exists when $r_{realized}$ declines with declining N, after reaching a maximum. (b) In the absence of an Allee effect, $r_{realized}$ continually increases as N becomes smaller, but the variance in $r_{realized}$, reflected by the thicker grey lines, might increase as N declines, increasing the uncertainty in $r_{realized}$ and, thus, increasing uncertainty in recovery.*

It is also important to appreciate that even if $r_{realized}$ does not decline with declining N in a deterministic sense (sub-section 6.3.4), Allee effects might still emerge from negative density dependence (i.e. $r_{realized}$ declining linearly with declining N). This can come about if there is an increase in the variance in $r_{realized}$ as population size declines (Figure 9.5(b)). Such a pattern of increased variance in $r_{realized}$ with declining population size has been documented in fishes (Minto et al. 2008). The greater the variance in $r_{realized}$, the greater the uncertainty in $r_{realized}$. This might be one reason why small populations are more vulnerable to environmental stochasticity, demographic stochasticity, and genetic stochasticity than large populations (Lande 1993).

9.6.3 The portfolio effect

Correlates of impaired recovery (Table 9.7) draw attention to the potential for within-population differences in life history to influence recovery. It is again useful (sub-section 2.4.5) to consider each life history as a solution that natural selection has produced to solve the problem of how to persist in a given environment. Multiple life histories within or among populations offer the possibility of multiple solutions to addressing predictable and unpredictable environmental changes, thus enabling populations or groups of populations to better resist disturbance.

An analogous set of circumstances exists in the financial world where individuals are often advised to invest in a diversified set of stocks. Doing so, in theory, allows individuals to better withstand unanticipated downturns in the stock market and not have their entire portfolio suffer as a consequence. In population dynamical terms, a diversified investment portfolio should increase the resistance of an investor, rendering them better able to weather decreasing stock prices in some of their investments.

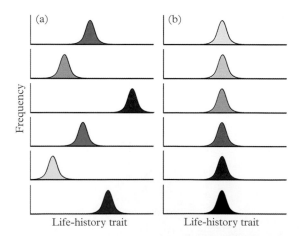

Figure 9.6 *The 'portfolio effect' among populations. A species comprised of (a) populations that differ in life history can potentially provide the species with greater resistance to environmental change and, thus, more stability over time than (b) a species comprised of populations that differ little in life history.*

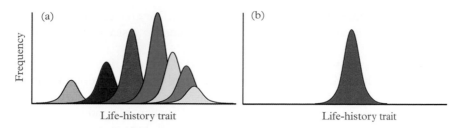

Figure 9.7 *The 'portfolio effect' within populations. A population comprised of (a) individuals that differ in life history can potentially provide the population with greater resistance to environmental change and, thus, more stability over time than (b) a population in which individuals express the same life history.*

An instructive example of this 'portfolio effect' in a life-history context has been offered as an explanation for the stability of catches in a fishery for sockeye salmon (*Oncorhynchus nerka*) in Bristol Bay, Alaska (Figure 9.6) (Schindler et al. 2010). The catch or 'success' of the fishery depends on contributions from different sockeye populations in different years. Some populations fare better in some years and provide a greater proportion of the yearly catch; others thrive in other years. The salmon populations differ in size and age at maturity. They also differ in the timing of migration from fresh water to and from the ocean, and in shape characteristics that render them more or less vulnerable to bear predation.

The concept behind the portfolio effect can also be applied within populations (Figure 9.7). It is reasonable to hypothesize that populations comprised of a breadth of variation in life-history traits might be better able to resist environmental change and recover more rapidly from depletion.

9.7 Summing Up and a Look Ahead

Human-induced disturbances can affect age-specific survival (l_x) and fecundity (b_x). This has potential to affect r_{max} with negative consequences for species viability and persistence. There are several types of assessments used to classify the vulnerability of species and populations to extinction, exploitation, and climate change. When information on r_{max} is unavailable, vulnerability assessments often rely on correlates of r_{max}. These have included generation time, age at maturity, maximum size, longevity, fecundity, natural mortality, and individual growth rate. Empirical research indicates that links with r_{max} are strong for some traits, such as age at maturity and body size, but weak for others, such as fecundity.

In addition to assessments of declining species, there have been efforts to identify factors that affect the rate and uncertainty of recovery. One idea to emerge from such work is the notion that the resistance of species to environmental change, and their resilience in recovering following depletion, is affected by the magnitude of life-history variation within and among populations. The greater the variability in life history within a population, the greater that population's resistance and resilience. The greater the life-history variability among populations within a species, the greater the resistance and resilience of that species.

In Chapter 9, links between l_x, b_x, and r_{max} provided a foundation for illustrating how a knowledge of life histories can inform assessments of species vulnerability and recovery. These links offer a logical and empirical basis for applying life-history theory to resource management, including factors affecting sustainable rates of exploitation, estimation of natural mortality, and the degree to which harvesting can generate life-history evolution. These topics are explored in Chapter 10.

10

Applications: Sustainable Exploitation of Evolving Resources

10.1 Sustainability

The concept of sustainability permeates government policies, regulatory frameworks, and laws. It is central to how many people view the extraction and use of renewable resources. In a global context, the aspiration of sustainable development was spearheaded by a 1987 UN report produced by the World Commission on Environment and Development. Chaired by Norway's former prime minister, Gro Harlem Brundtland, this report, entitled *Our Common Future* (often called the Brundtland Report), articulated global challenges associated with sustainable use of human and natural resources. Two of these challenges addressed topics touched on in this book: species extinction and sustainable exploitation of wild organisms. The former was highlighted in Chapter 9. The latter will be explored further here.

Many of the challenges pertain to fisheries, a key source of protein for much of the world and the only global enterprise that captures undomesticated wild organisms directly (seafood, nutraceuticals) and indirectly (plant fertilizers, livestock feed) for human consumption. But fisheries are not the only source of protein obtained from wild populations. Many terrestrial animals are hunted for food, including bushmeat, often harvested in tropical and subtropical regions. Governments and regulatory agencies recognize multiple axes of sustainability, including socio-economic, cultural, and ecosystem objectives (Stephenson et al. 2019). But no matter how influential these objectives, they are ultimately constrained by the bounds of sustainability generated by organismal ecology, evolution, and life history.

Exploitation directly affects age-specific survival probabilities. It can also affect age-specific fecundity schedules, particularly among indeterminately growing species for which exploitation causes density-dependent changes in individual growth (such that a reduction in population density leads to a phenotypically plastic increase in individual growth because of reduced competition for resources). Life-history changes generated by fishing and hunting affect $r_{realized}$, often resulting in shifts in population size, diminished breadth in age and size structure, and in some cases increased susceptibility to natural sources of mortality, such as that resulting from predation.

A Primer of Life Histories: Ecology, Evolution, and Application. Jeffrey A. Hutchings, Oxford University Press. © Jeffrey A. Hutchings 2021.
DOI: 10.1093/oso/9780198839873.003.0010

To achieve sustainable rates of exploitation, science has increasingly accounted for species and population differences in life history. As the title of this chapter serves to remind, exploited populations are evolving entities. This means that, in addition to natural selection, they can be subjected to harvest-induced selection imposed by fishing and hunting.

A tangible manifestation of the seeds sown by the Brundtland Report are the Sustainable Development Goals (SDGs) adopted by UN member states in 2015. Of the 17 SDGs, Life Below Water (SDG 14) is most clearly linked to exploitation: 'Conserve and sustainably use the oceans, sea and marine resources for sustainable development'. This SDG was motivated primarily by multiple fishery collapses spanning the globe from the 1960s through the 1990s. Foremost among these were fishing-induced depletions of Atlantic cod.

10.2   The Collapse of Canadian Cod

Managed today as more than 20 units or 'stocks' throughout the North Atlantic, the once most numerous stock of Atlantic cod—Northern cod—inhabits waters from southern Labrador to the northern half of the Grand Banks off eastern Newfoundland. Commercially fished by the Portuguese since at least the 1470s (Cole 1990), by the mid-1500s several hundred vessels sailed annually in spring from Portugal, Spain, France, and England to fish these waters, returning home again in autumn (Castañeda et al. 2020). Their primary focus was Northern cod (Figure 10.1).

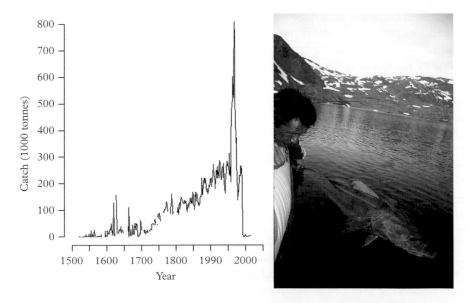

Figure 10.1 *Left panel shows a catch reconstruction of Northern cod from 1508 to 2018 (updated from Hutchings and Myers 1995). Right panel was taken from Ogac Lake, Baffin Island, Nunavut, Canada, in July 2003 during field research by David Hardie, Michael Mipeegaq, and Jeffrey Hutchings.*

Photo © David Hardie, reprinted with permission by the photographer.

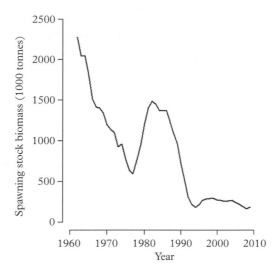

Figure 10.2 *Based on estimates of the spawning stock biomass of all Canadian cod combined, the total weight of breeding individuals declined by more than 2 million tonnes (Hutchings and Rangeley 2011).*

Having been fished sustainably for centuries at annual levels of less than ∼ 250 000 tonnes, reported catches of Northern cod (which exclude discarded and misreported fish) spurred by new and unregulated technology (factory-freezer trawlers) exceeded 800 000 tonnes in 1968 (Hutchings and Myers 1995). Thereafter, Northern cod declined and so did catches, for which annual quotas were not legally enforced until the late 1970s. A commercial fishing moratorium was announced by the Canadian government in 1992. By 1994, all remaining cod fisheries in Canada were closed. As of 2021, the population sizes of all Canadian cod stocks were smaller than their biomass 'limits' set by fisheries management under the auspices of the Precautionary Approach (section 10.4), meaning they are in a critical state, according to Canadian sustainable fisheries policy.

To place the decline of the species in a non-fisheries context, the collapse of Atlantic cod represents the greatest numerical loss of a vertebrate in Canada. The reproductive component of all Canadian cod combined declined more than 90 per cent between the early 1960s and the early 1990s (Figure 10.2) and, for all intents and purposes, remains at the same depressed level today. Numerically, this was a reduction of between 1.5 and 2.5 billion breeding individuals. By weight, this is roughly equivalent to 27 million humans (Hutchings and Rangeley 2011).

10.3   Maximum Sustainable Yield

10.3.1   The basic concept

Populations of Atlantic cod were depleted because they were exploited at unsustainable rates of fishing. A simple means of assessing whether a population is being sustainably harvested is to compare the current harvest with the maximum that the population can theoretically sustain for the foreseeable future. First articulated by Russell (1931), the concept of maximum sustainable yield (MSY) is today embedded in jurisdictional tools for achieving fisheries sustainability, such as New Zealand's *Fisheries Act*, the European Common Fisheries Policy, and the UN's Sustainable Development Goal 14 (Life Below Water).

Although this was not necessarily true in the late twentieth century, when many global fishery collapses occurred, the most sustainable harvests today tend to be those blessed with the greatest amount of data. The gold standard would be information on the numbers of individuals at each age, the natural and harvest-induced probabilities of surviving from one age to the next, and the numbers of offspring produced by the average individual at each age. Based on the terms introduced in Chapter 1, and used throughout this book, these data can be recognized as key components of population demography (N_x) and life history (l_x, b_x), both of which influence key entities of sustainability: MSY and r_{max}. Another important consideration when evaluating population status is that the longer the time period over which these data are available, the greater the confidence that natural variability within each data set, such as periods of high and low productivity, has been captured.

Not surprisingly, these gold standards exist for few species or populations. This makes it challenging to determine harvest or catch levels that are sustainable, i.e. able to be maintained at the same levels for the foreseeable future. As discussed later in this chapter, various methods, based on life-history traits, have been proposed to address data deficiencies when assessing sustainability.

Irrespective of the methods used to estimate sustainable harvests, the premise that there exists a population size at which $\partial N/\partial t$ is maximized underlies most assessments of the sustainable exploitation of animals. These are ultimately based on the simple, dome-shaped density-dependent population growth model presented in sub-section 1.2.3. Reproduced here as Figure 10.3, it represents the relationship between population growth rate ($\partial N/\partial t$) and population size (N).

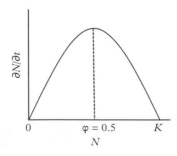

Figure 10.3 *In the simplest model of density-dependent population growth, the highest sustainable rate of production of new individuals ($\partial N/\partial t$) is obtained when the population size is half of carrying capacity, K. Thus, $\partial N/\partial t$ is maximized when $\varphi = 0.5$.*

The aim from a sustainable-harvesting perspective is to manage a population so that it is achieving the highest sustainable rate of production of new individuals over a particular time period (usually each year). The proportion of the carrying capacity (K) at which $\partial N/\partial t$ is maximized, and at which MSY can be harvested, can be represented by φ. For the simplest scenario (Figure 10.3), $\varphi = 0.5$.

10.3.2 MSY depends on population size

In reality, the relationship between $\partial N/\partial t$ and N need not be bilaterally symmetrical, leading to alternative dome-shaped relationships within populations. For example, models suggest that φ might be less than 0.5 for many marine fishes (Punt et al. 2014) but greater than 0.5 for marine mammals (Fowler 1988).

The concept of MSY is not restricted to fishing. It underscores, for example, efforts to assess the sustainability of subsistence and commercial hunts for bushmeat. Based on Cole's (1954) pioneering life-history work, Robinson and Redford (1991) proposed that the maximum sustainable production (P_{max}) of a population hunted for bushmeat can be expressed as:

$$P_{max} = 0.6K(r_{max} - 1)MF \qquad \text{Equation 10.1}$$

where K is carrying capacity and MF ('mortality factor') is a proportion that varies with longevity. For long-lived (>10 yr) species, MF is set to 0.2; for short-lived species, $MF = 0.6$. It is intended to reflect the idea that the shorter the lifespan, the greater the species' annual natural mortality rate, meaning that a higher proportion of the population will die due to natural causes and can, in theory, be hunted instead. The '0.6' in Equation 10.1 reflects the assumption that $\partial N/\partial t$ is maximized when the population is at 60 per cent of its carrying capacity (i.e. $\varphi = 0.6$), an assumption considered hypothetically reasonable for forest ungulates (Milner-Gulland and Akçakaya 2001).

Robinson and Redford's (1991) approach has been widely used in bushmeat sustainability assessments (Weinbaum et al. 2013). The strength of the approach lies in its computational simplicity. To estimate r_{max}, Robinson and Redford (1991) used the Euler-Lotka equation (Equation 5.3 in sub-section 5.3.3), thus incorporating information on age at maturity and fecundity. One rather serious weakness, however, is that the model excludes age-specific survival. Additional drawbacks lie in the difficulty in empirically estimating K and r_{max}, the arbitrariness of the mortality factor MF, and the assumption that $\varphi = 0.6$.

Based on life-history theory, Fowler (1988) came up with a clever means of estimating φ. He showed that φ is directly related to r_{max} when maximum per capita population growth is expressed as a function of generation time (G). He collated data on 16 species from which he was able to estimate φ, r_{max}, and G (Figure 10.4). The phylogenetic breadth of the data ranges from bacteria to blue whales (*Balaenoptera musculus*). The value of φ declines linearly and predictably as the natural logarithm of $r_{max}G$ increases, such that the relative population size at which $\partial N/\partial t$ is maximized declines as $r_{max}G$ increases. Since Fowler's (1988) initial work, $r_{max}G$ has been used to identify overharvested bird

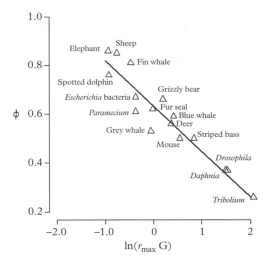

Figure 10.4 *Relationship between φ (the proportion of carrying capacity at which ∂N/∂t is at its maximum) and $r_{max}G$ (maximum per capita growth rate multiplied by generation time). The line corresponds to the equation φ = 0.633 − 0.187 × ln($r_{max}G$).*

Redrawn from Fowler (1988) with data provided by Charnov (1993) and reprinted by permission from Springer Nature.

populations (Niel and LeBreton 2005) and to evaluate the sustainability of fishery bycatches of seabirds and sharks (Dillingham et al. 2016).

10.3.3   Estimating MSY

There are various means of estimating sustainable yields. Robertson and Redford's (1991) approach for estimating MSY in bushmeat hunting, notwithstanding its drawbacks, was ultimately based on estimates of r_{max} and K. The same is true of more operationally complex models used in fisheries stock assessments. The utility of these models depends on the availability and reliability of catch data and model parameter estimates.

Perhaps the simplest method (simplest in interpretation) for estimating MSY is the 'surplus production model' (Hilborn 2001; Sparholt et al. 2020). This model is founded on the classic density-dependent model of population growth (Figure 10.3), expressed in Chapter 1 as Equation 1.14, for which population growth rate (∂N/∂t) is equal to:

$$\partial N/\partial t = r_{max} N_t (1 - N_t/K) \qquad \text{Equation 10.2}$$

Thus, the change in population size from time step t to $t + 1$ (one year in fisheries models), or $N_{t+1} - N_t$, can be expressed as the population size at time t plus the number of new individuals added to the population because of its growth, such that:

$$N_{t+1} = N_t + r_{max} N_t (1 - N_t/K) \qquad \text{Equation 10.3}$$

In fisheries, rather than numbers of individuals, the size of a population (or the managed 'stock') is usually expressed in terms of weight or biomass (B), such that Equation 10.3 becomes:

$$B_{t+1} = B_t + r_{\max} B_t (1 - B_t / K) \qquad \text{Equation 10.4}$$

For a population subjected to fishing, growth is tempered by the number of fish removed by the harvest, i.e. the catch (C). This means that in an exploited population,

$$B_{t+1} = B_t + r_{\max} B_t (1 - B_t / K) - C_t \qquad \text{Equation 10.5}$$

When a population is stable and at equilibrium, $B_{t+1} = B_t$ (meaning that $B_{t+1} - B_t = 0$), and the annual equilibrium catch is:

$$C_t = r_{\max} B_t (1 - B_t / K) \qquad \text{Equation 10.6}$$

or:

$$C_t = r_{\max} B_t - r_{\max} B_t^2 / K \qquad \text{Equation 10.7}$$

Equation 10.7 is that of a parabola. Using basic calculus, the maximum of the parabola, which corresponds to the value of B_t at which C_t is maximal (i.e. B_{MSY}), can be found by taking the first derivative of Equation 10.7 with respect to B_t, setting it to zero, and solving for B_t. MSY can be calculated by inserting B_{MSY} into Equation 10.7 and solving for C_t.

In a fisheries context, as in the bushmeat example given in Equation 10.1, the rate of growth (i.e. $r_{\max} B_t (1 - B_t/K)$) is termed 'production' (P). Equation 10.5 thus becomes simplified to read:

$$B_{t+1} = B_t + P_t - C_t \qquad \text{Equation 10.8}$$

A semantic twist to this equation is that P can be termed surplus production. If a population remains stable through time, such that $B_{t+1} = B_t$, then the production of new individuals every year, P, over and above that required to maintain a stable population size can be thought of as being 'surplus' and, thus, available to be caught. Re-arranging Equation 10.8 by placing surplus production on the left side yields:

$$P_t = B_{t+1} - B_t + C_t \qquad \text{Equation 10.9}$$

This is the surplus production model (Hilborn 2001; Figure 10.5(a)). MSY is obtained at the population size at which surplus production (analogous to $\partial N/\partial t$; Chapter 1) is highest. This population size is termed B_{MSY} (or SSB_{MSY}, the spawning stock biomass that yields MSY). In a fisheries context, carrying capacity is often termed B_0 (also called 'virgin' or 'unfished biomass').

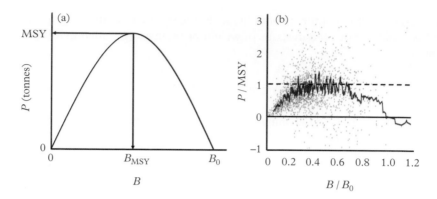

Figure 10.5 *Surplus production model. (a) Relationship between surplus production (P) and population biomass (B) is dome-shaped, extending from zero to the unfished biomass, B_0. The biomass that produces the maximum sustainable yield (MSY) is B_{MSY}. (b) A compilation of data for 53 marine fish populations for which MSY (where P/MSY = 1) is achieved at values of B/B_0 (i.e. φ) ranging from roughly 0.3 to 0.6. The red line is a running mean of 25 points.*

Panel (b) is re-drawn from Sparholt et al. (2020) and reprinted by permission from Oxford University Press.

Combining multiple time-series data for 53 data-rich marine fish populations, Sparholt et al. (2020) found that surplus production (relative to MSY) is related to relative population size (B/K) by a roughly dome-shaped curve. This suggests that surplus production models are able to capture the underlying production dynamics of fish populations. The data compilation indicates that MSY is not always achieved at $0.5K$, but that φ can vary between 0.3 and 0.6 (Figure 10.5(b)). This is reflected by the values of B/B_0 that encompass ratios where $P/MSY = 1$, i.e. where the surplus production is equal to the maximum sustainable yield.

10.3.4   MSY for data-poor species

The idealized output of the surplus production model (Figure 10.5(a)) might suggest that estimation of MSY is relatively straightforward. It can be, but it depends a great deal on the quality of the underlying data used to generate P and B and on the assumptions that r and K are temporally stable, all individuals are identical, and the response to density dependence is immediate. One also requires sufficient contrasts in the data, ideally including observations of B at or near virgin (unfished) stock sizes. For data-poor species, these data do not exist, a reality that has necessitated the use of simpler methods to estimate MSY, such as the one introduced by Robinson and Redford (1991) for bushmeat (Equation 10.1).

One example of such a method for data-poor fisheries, termed the 'Catch-MSY' or CMSY method (Froese et al. 2017), uses data on catch and estimates of population size (B) to estimate r_{max}, K, and MSY. Schijns et al. (2021) applied the CMSY approach to the five centuries of catch estimates for Northern cod shown in Figure 10.1, perhaps

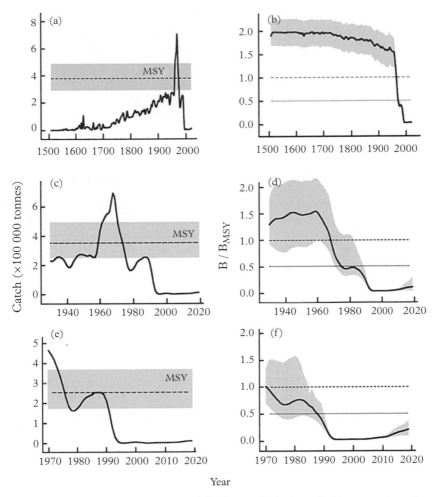

Figure 10.6 *Catch and population biomass of Northern cod relative to the biomass estimated to produce maximum sustainable yield (B$_{MSY}$): (a, b) 1508 to 2019; (c, d) 1930 to 2019; and (e, f) 1970 to 2019. The catch and relative biomass level (B/B$_{MSY}$) compatible with MSY are shown (dotted lines) along with the 95% confidence intervals (shaded regions).*

From Schijns et al. (2021).

the longest time series of catches for any fishery in the world. Using Monte-Carlo and Bayesian state-space modelling (methods beyond the scope of this primer), their estimates of maximum per capita population growth and carrying capacity (± 95 per cent confidence intervals) were, respectively, $r_{max} = 0.25$ (± 0.14 – 0.46) and $K = 6.0$ (± 4.0 – 8.9) million tonnes (Figure 10.6). Their estimate of MSY, again based on the entire time series (1508–2019), was ~ 380,000 tonnes per year (± 290,000 – 490,000). The population biomass is estimated to have fallen below the biomass capable of producing MSY in the 1960s.

In contrast to fisheries, the harvest quotas of hunts for terrestrial mammals such as moose (*Alces alces*), elk, deer, bears, and bobcats are often not based on quantitative assessments of MSY (Fryxell et al. 2010). This is not because MSY cannot be estimated for terrestrial species or that the theoretical basis for MSY does not apply to animals on land. Fryxell et al. (2001), for example, reported a twenty-year sustainable harvest rate of martens (*Martes americana*) in central Canada that was just less (34 per cent) than the estimated MSY for the population (36 per cent). Rather, many hunts are not regulated on the basis of achieving MSY because of a lack of relevant data.

10.3.5   Understanding mortality: the key to sustainability

A pervasive theme throughout this book is the fundamental consideration that to understand life-history evolution one needs to understand how l_x changes with age. This is also true when trying to achieve rates of sustainable harvesting. As a general rule, l_x declines with age in the curvilinear manner depicted in Figure 10.7. In this particular example, from an initial 100 offspring, there are 67 individuals alive at age 1 and 45 individuals alive at age 2, meaning that $l_1 = 0.67$ and $l_2 = 0.45$.

The line in Figure 10.7 corresponds to the following equation:

$$N_x = N_0 e^{-Zx} \qquad\qquad \text{Equation 10.10}$$

where the number of individuals alive at age x (N_x) is a function of N_0 (number of offspring) and Z (instantaneous rate of mortality). Mortality expressed as a proportion is calculated as $1 - e^{-Z}$ and the proportion surviving is e^{-Z}. Thus, in Figure 10.7, where $Z = 0.4$, the mortality and survival expressed as proportions are 0.33 and 0.67, respectively.

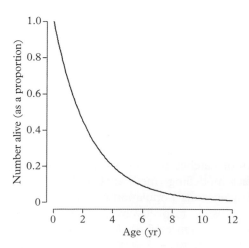

Figure 10.7 *From an initial cohort or year-class of 100 offspring, the number of individuals alive at subsequent ages (x) declines. In this example, age-specific survival, l_x, is the number of individuals alive at age x divided by 100. The line in this figure corresponds to $N_x = N_0 e^{-Zx}$, where $N_0 = 100$ and $Z = 0.4$.*

For an exploited population, Z must be partitioned into the instantaneous rates of mortality attributable to natural causes (M, or natural mortality) and exploitation (F, or exploitation mortality, termed fishing mortality in the fisheries literature). Thus, Equation 10.10 becomes:

$$N_x = N_0 e^{-(M+F)x}$$ Equation 10.11

The catch (C) divided by the size of the harvestable portion of the population, $N_{harvest}$ or $B_{harvest}$ (not all individuals in a population will be vulnerable to a fishery or a hunt), is the harvest rate (HR), i.e. $C/N_{harvest}$. The fishing mortality, F, can then be calculated as:

$$F = -\ln(1 - HR).$$ Equation 10.12

For example, if the catch is 300 and $N_{harvest}$ is 1000, the harvest rate is $300/1000 = 0.30$, and F is $-\ln(1 - 0.30) = 0.36$.

The parameter that is the most challenging to estimate for any wild population is natural mortality, M. A mark-recapture study (White and Burnham 1999) can be used to estimate M if the size of the marked (i.e. tagged) population is known, catch is known, and estimates of tag-reporting rate and tag loss are available. In the absence of exploitation, M can also be estimated from fisheries independent surveys. However, mark-recapture experiments are not always feasible and fisheries independent survey data are not available for many populations.

This has led to the use of life-history traits to estimate M, particularly for data-poor species. For fishes, estimates of M are often based on metrics of maximum age, maximum body size, and individual growth (Table 10.1). Although based solely on life-history attributes, some estimates of M, such as those calculated from expressions (1) to (3) in Table 10.1, have been shown to be strongly correlated with direct, independently made estimates of M in natural populations (Gislason et al. 2010; Then et al. 2015; Waples and Audzijonyte 2016). For a number of animals, estimates of M can be based on life-history invariants, using age at maturity, the von Bertalanffy growth coefficient, or clutch size (expressions (4)–(13)).

Table 10.1 *Thirteen expressions used to estimate natural mortality, M, as a function of life history: t_{max} = maximum age; k = von Bertalanffy growth coefficient; L_∞ = von Bertalanffy asymptotic length; L_α = length at maturity; c_1 = a parameter ranging between 1.65 and 2.20; α = age at maturity; c_2 = a parameter ranging between 1.65 and 2.10; b = yearly clutch size, in daughters. Notes on taxon identifiers: 'shrimp' are species from the family Pandalidae; 'sea urchins' are in the class Echinoidea; 'reptiles' include snakes and lizards; 'mammals' include terrestrial species only. All but the first three expressions are examples of life-history invariants (sub-section 2.3.2).*

Expression	Taxon	Reference
(1) $M = 4.99\ t_{max}^{-0.916}$	fishes	Then et al. (2015), updated from Hoenig (1983)
(2) $M = 4.11\ k^{0.73}\ L_\infty^{-0.331}$	fishes	Then et al. (2015), updated from Pauly (1980)

(Continued)

Table 10.1   Continued

Expression	Taxon	Reference
(3) $M_\alpha = (L_\alpha/L_\infty)^{-1.5} \times k$	fishes	Charnov et al. (2013), updated from Gislason et al. (2010)
(4) $M = c_1/\alpha$	fishes	Beverton and Holt (1959); Charnov (1993); Jensen (1996)
(5) $M = 2.10/\alpha$	shrimp	Charnov (1993)
(6) $M = 1.32/\alpha$	reptiles	Charnov (1993)
(7) $M = 0.73/\alpha$	mammals	Charnov (1993)
(8) $M = 0.79/\alpha$	primates	Charnov (1993)
(9) $M = c_2\,k$	fishes	Charnov (1993); but see Figure 2.15 (Thorson et al. 2017)
(10) $M = 1.5\,k$	reptiles	Charnov (1993)
(11) $M = k$	sea urchins	Charnov (1993)
(12) $M = 2.56\,k$	shrimp	Charnov (1993)
(13) $M = 0.2\,b$	birds	Charnov (1993)

10.4   Reference Points to Guide Sustainability Initiatives

10.4.1   Limits, targets, and MSY

The Precautionary Approach is the guiding principle for the sustainable exploitation of wild animals. Its purpose is to protect the environment (such as an exploited population) and limit risk (such as population collapse) by taking preventive action in response to threats of harm, including situations of scientific uncertainty.

In a fisheries context, application of the Precautionary Approach was agreed to by signatories to the UN Fish Stocks Agreement in 1995. A key element of the Agreement is its guidance for applying precautionary reference points. Conservation or limit reference points should be established to constrain harvesting within safe biological limits. Target reference points are set to identify goals to meet sustainable harvesting objectives. Many fishery jurisdictions have established reference points for fishing mortality (F) and population biomass (B). Generic representations of these are illustrated in Figure 10.8.

It is common for reference points to be set in relation to MSY. Some jurisdictions set B_{MSY} as the target population biomass, although increasingly many are setting B_{target} at a higher level. In Australia, for example, the default for B_{target} is 1.2 B_{MSY} (Australia 2018). Setting B_{target} at a level higher than B_{MSY} is perceived as being precautionary by accounting

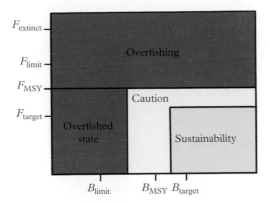

Figure 10.8 *A generic illustration of how reference points for population biomass (B) and fishing mortality (F) can be used to qualitatively assess the health of a fish stock. Reference points for F and B specify targets, limits, levels associated with maximum sustainable yield (MSY) and, for F, extinction.*

for uncertainties in estimating B_{MSY} and by reducing the probability that fishing will push the biomass below B_{MSY}.

The biomass limit reference point, B_{limit}, set well below B_{MSY}, is intended to demarcate a threshold below which the population is likely to experience serious harm (e.g. severely reduced reproductive capacity, impaired recovery potential, increased probability of extinction). In Canada, the provisional default setting for B_{limit} is 0.4 B_{MSY} (Fisheries and Oceans Canada 2009); in Australia, the minimum standard is that B_{limit} should be equal to or greater than 0.5 B_{MSY} (Australia 2018). It is common for fisheries to be closed if population biomass falls below B_{limit}.

Similar reasoning applies to reference points for fishing mortality. Although F_{MSY} was once considered a target, it is now more commonly viewed as a limit (e.g. Fisheries and Oceans Canada 2009; Australia 2018). Correspondingly, the target fishing mortality, F_{target}, is often set below F_{MSY} to err on the side of precaution. Figure 10.8 also identifies $F_{extinct}$, the fishing mortality, if exerted over a sufficiently long period of time, that will drive a population to extinction. To a first approximation, $F_{extinct} = r_{max}$ (Dulvy et al. 2004). A related metric, called F_{crash}, has been estimated as 2 F_{MSY} (Zhou et al. 2009).

The state space created by Figure 10.8 provides a template that allows for the sustainability of a fishery (in principle, any harvested aquatic or terrestrial population) to be deduced and reported. Overfishing occurs when F is too high; populations are in an overfished state if B is too low. A zone of caution, whose relation to reference points varies among jurisdictions, separates the state space represented by overfishing and overfished states from levels of B and F that denote sustainability.

Although Figure 10.8 exudes clarity, the devil is in the details. In practice, there can be substantive differences in how jurisdictions set reference points and how they define different states of sustainability, creating challenges for those interested in comparing sustainability among fisheries and among countries. One example is the level of B that constitutes a population in an overfished state (Table 10.2).

Table 10.2 *Definitions of what constitutes an overfished population in various jurisdictions, expressed in relation to B$_{MSY}$ (population biomass at which maximum sustainable yield is achieved; also expressed as SSB$_{MSY}$ for spawning stock biomass) and B$_{limit}$ (population size below which reproductive productivity is impaired). B$_0$ is analogous to carrying capacity. Definitions from Hilborn (2020) and FAO (2018).*

Management Jurisdiction	Definition of Overfished
International Council for the Exploration of the Sea (European fisheries); Japan	$B < B_{limit}$
US National Marine Fisheries Service	$B < 0.5 \, B_{MSY}$
New Zealand	$B < 0.5 \, B_{MSY}$ or $B < 0.2 \, B_0$
Australia	$B < B_{limit}$, where $B_{limit} = 0.2 \, B_0$
Canada	$SSB < 0.8 \, SSB_{MSY}$
Chile; Tuna Commissions (Atlantic, Western and Central Pacific, Inter-American Tropics)	$SSB < SSB_{MSY}$
Indian Ocean Tuna Commission	$SSB < SSB_{unfished}$
Commission for the Conservation of Southern Bluefin Tunas	No definition
Food and Agriculture Organization of the United Nations (FAO)	$B < B_{MSY}$

10.4.2   Reference points and life history

There is an obvious implicit relationship between reference points and life histories. By and large, reference points are set in relation to MSY. B_{MSY} depends on the relative population size at which population growth rate (or surplus production) is maximized, i.e. φ. As Fowler (1988) and others have shown, φ depends on generation time and r_{max}, both of which ultimately depend on age-specific survival (l_x) and fecundity (b_x). MSY and F_{MSY} are also both based on r_{max}. But, as noted previously, r_{max} can be difficult to estimate in practice because doing so requires more data than what is available for many populations.

Le Quesne and Jennings (2012) offered an illuminating solution to the challenge of setting reference points for data-poor species. The analytically appealing element to their work is that their population models are ultimately based only on observations of maximum body size (L_{max}). They then apply relationships between L_{max} and other life-history traits, including life-history invariants, to parameterize age-structured models which are then used to estimate reference points for fisheries' sustainability and conservation (Figure 10.9).

To know the sensitivity of a species to fishing, it is necessary to identify the levels of fishing mortality (F) it can sustain. The approach taken by Le Quesne and Jennings (2012) has three steps. Firstly, identify the species and their maximum body sizes. Secondly, develop an age-structured population model based on life-history relationships and invariants to determine reference points from a fisheries perspective (based on

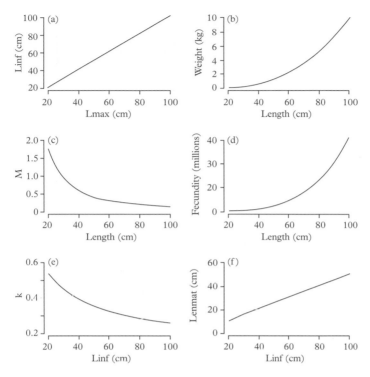

Figure 10.9 *Life-history relationships used by Le Quesne and Jennings (2012) to parameterize an age-structured population model. Linf = von Bertalanffy asymptotic length (L_∞); M = natural mortality; k = von Bertalanffy growth coefficient; Lenmat = length at maturity. Equations: (a) $\log_{10}(L_\infty) = 0.044 + 0.9841 \times \log_{10}(L_{max})$; (b) $W = 10^{-5} \times L_t^3$; (c) $M = exp(0.55 - 1.61 \times ln(L_t) + 1.44 \times ln(L_\infty) + ln(k))$, where $L_\infty = 134$ and $k = 0.11$ (Northeast Arctic cod; Wang et al. 2014); (d) Fecundity $= 10^{-6} \times L_t^{3.75}$; (e) $k = 2.15 \times L_\infty^{-0.46}$; (f) Lenmat $= 0.64 \times L_\infty^{0.95}$.*

proxies of MSY) and a conservation perspective (based on spawning stock biomass and reproductive output). Thirdly, assess species sensitivity to fishing by comparing these reference points to actual and potential levels of fishing mortality, *F*.

For 124 demersal fish species in the Celtic Sea, Le Quesne and Jennings (2012) concluded that conservation reference points, which can be thought of as limit reference points, ranged between $F = 0.05$ for two species of skate (elasmobranch fishes) to >1 for teleosts (bony fishes) having a maximum body size smaller than 38 cm. If *F* exceeds a species conservation reference point for *F*, the conclusion would be that fishing pressure had potential to drive population sizes to depleted levels sufficient to threaten their viability.

Figure 10.10 shows how the proportion of assessed species at risk of extirpation changes with fishing mortality, *F*. It also shows the 2020 levels of *F* for two species targeted by fishing in the Celtic Sea: cod (ICES 2020a) and hake (ICES 2020b). Assuming that species caught as bycatch experience the same levels of *F* as the targeted species, at

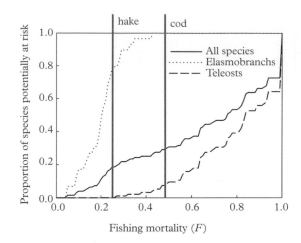

Figure 10.10 *The proportion of species of marine fishes in the Celtic Sea (n = 124) assessed as being at risk, as a function of fishing mortality (F). Data for elasmobranchs (sharks, skates, rays) are distinguished from teleosts (bony fishes). The F experienced by hake* (Merluccius merluccius) *and cod* (Gadus morhua) *in 2020 are indicated by blue vertical lines.*

Reprinted by permission from John Wiley and Sons

the F applied to hake in 2020 (0.26), more than 79 per cent of elasmobranchs in the Celtic Sea are potentially fished at rates higher than their conservation limit reference point. At the 2020 fishing mortality applied to cod ($F = 0.48$), all elasmobranchs and five to ten per cent of teleosts are potentially threatened.

Being mindful of the model assumptions, the elegant simplicity of Le Quesne and Jennings' (2012) method is that it only requires knowledge of taxonomy, body size, and knowledge of how body size is related to life-history traits, potentially allowing for rapid assessment of sensitivity to fishing for data-poor fisheries around the world.

10.5 Harvest-Induced Evolution

10.5.1 Fisheries-induced changes in life history

Thus far, this chapter has focused on how life histories affect the sustainability of exploitation. Here, we explore how exploitation affects life history. Consider a previously unexploited population at or near carrying capacity. Fishing or hunting pressure exerted on that population will result in decline. From an unexploited state, to achieve MSY, managers might wish to reduce populations by 40 per cent or more, depending on the population or species (sub-section 10.3.2). According to the principles of density-dependent population regulation, a reduction in abundance or density will increase $r_{realized}$ (sub-section 1.2.3). Although in theory this increase might be a consequence of increases in

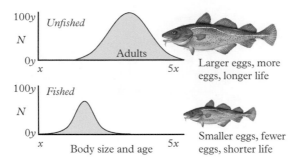

Figure 10.11 *Fishing reduces population size (N). It can also shift and narrow a population's age and size distribution with life-history consequences for adults.*

Line drawing of Atlantic cod © H.L. Todd.

either l_x or b_x, in reality $r_{realized}$ increases primarily because of increased juvenile and/or adult survival (sub-section 5.5.4).

Proximate reasons for increased survival include reduced competition for, and thus greater access to, limiting resources such as food. For ectotherms, greater access to food leads to faster growth and, all else being equal, faster growth leads to earlier age at maturity (Wootton 1998; Roff 2002). In other words, a fishing-induced reduction in density or abundance can be expected to result in individuals attaining maturity at a younger age when the population is subjected to fishing, compared to individuals in the same population in an unfished state. Fishing often has other life-history consequences, including a narrowing or truncation in the age and size distribution (Barnett et al. 2017) coupled with smaller size at maturity, leading to the average female producing fewer eggs, smaller eggs, and experiencing a shorter life (Figure 10.11).

Earlier maturity can be readily explained as a phenotypically plastic response (section 3.3) to faster growth, insofar as faster growth reduces the time required to attain a body size at which reproduction is physiologically and developmentally possible. However, earlier maturity in exploited populations is not always associated with faster growth (Trippel 1995; Hutchings 2005). In some fish populations, individual growth rate has been observed to decline as fishing mortality increased and population size declined (cod in the Southern Gulf of St. Lawrence, Canada; Swain et al. 2007). In addition to changes in age at maturity, size at maturity has also been observed to decline in association with intense fishing pressure. Median length at maturity among Eastern Scotian Shelf cod, for example, declined from approximately 42 cm in the late 1970s to 32 cm in the early 2000s (Hutchings 2005). Fishing-associated reductions in age and length at maturity of 20 to 30 per cent are not uncommon when examining species worldwide (e.g. Hutchings and Baum 2005; Sharpe and Hendry 2009).

Although reductions in age at maturity can be explained as plastic responses to reduced abundance, reductions in size at maturity and individual growth are less easily interpretable. This raises the question as to whether life-history changes might be caused

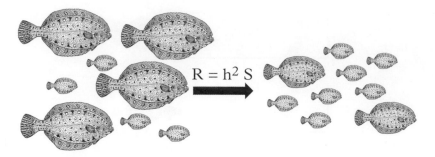

Figure 10.12 *Fishing has potential to generate evolutionary change in fished populations. One potential genetic shift in life history is evolution towards earlier age and smaller size at maturity. To a first approximation, if information was available on the heritability of the life-history trait(s) (h^2) and the selection differential (S), the response to selection (R) could be quantified (sub-section 3.2.2).*

by changes in gene frequency. By selecting against individuals whose genes predispose them to breed at older ages and larger sizes, fishing might favour genotypes that mature at relatively young ages, at small body sizes, or that grow at comparatively slow rates (Stokes and Law 2000; Law 2007; Swain et al. 2007) (Figure 10.12).

10.5.2   Fisheries-induced evolution

The potential for fisheries to alter life history was recognized in the early twentieth century. Cloudsley Rutter, a salmon biologist in California, offered the opinion in 1902 that the body size of salmon would decline if fisheries targeted only the largest individuals, thus increasing the preponderance of smaller individuals in future breeding populations (Dunlop et al. 2009). The geneticist J.B.S. Haldane explicitly identified fishing as a reason for 'observable evolution' proceeding with 'extreme and abnormal speed' (Haldane 1937: 338).

Handford et al. (1977) published the first empirically defensible examination of whether fishing could generate genetic change in exploited populations, studying a gill-net fishery for lake whitefish (*Coregonus clupeaformis*) in Alberta, Canada. This study was followed by Ricker's (1981) analysis of fishery-associated declines in average body size in Pacific salmon and Stokes et al.'s (1993) volume on harvest-induced evolution in animal populations. By the early 2000s, the development of probabilistic maturation reaction norms by Heino et al. (2002) opened up the possibility that genetically based phenotypic shifts in life-history traits could be detected by a method thought to disentangle growth-related phenotypic plasticity from genetic change (Heino et al. 2015).

The hypothesis that predators can generate evolutionary change in their prey has a long and rich history (Michod 1979; Endler 1986; Abrams 2000). Given the very high mortalities that fishing has exerted on wild populations (*F* can often be more than two to three times higher than *M*), it is logically defensible to hypothesize that some of the changes in life history associated with intense fishing, or hunting, can be attributable to evolution. This process is known as harvest-induced or fisheries-induced evolution (HIE and FIE, respectively).

One of the most illuminating selection experiments on the evolutionary consequences of harvesting used clonal mixtures of the zooplankton *Daphnia magna*. Edley and Law (1988) culled either small or large individuals from the experimental populations. Clones selected by culling small-sized individuals grew rapidly through the small size classes, maturing at a relatively older age. Clones selected by culling large-sized individuals (selection similar to that imposed by most commercial fisheries) responded by growing slower, maturing at an earlier age and, at some ages, by increasing their reproductive effort. Richard Law briefly described these results and their potential importance in a letter he wrote to me on 31 October 1989:

> I'm enclosing some papers on life histories and fisheries. One of the fisheries is for *Daphnia* in bottles, and it's a big jump from this to the real thing, but it does at least show that life histories can evolve under fishing and that this feeds back to the yield the population can sustain. The most recent manuscript (in press in Evolutionary Ecology) puts forward an argument that we should be able to control the evolution of harvested populations by the pattern of mortality we apply to them—I guess this is going to be controversial.

Perhaps the most compelling evidence of fisheries-induced evolution in a wild population is for Atlantic cod in Canada's Southern Gulf of St. Lawrence. Swain et al. (2007) used a 30-year time series of data to explore the question of whether decreased rates of individual growth could be explained as genetic responses to fishing, independently of the effects of population density and water temperature on growth. They analysed the ear bones, or otoliths, which reveal information on size-at-age throughout an individual's life. When the otoliths are sectioned, annuli are visible that represent years (much like tree rings). And, like tree rings, the length that an individual was at previous ages can be back-calculated from each otolith. Swain et al. (2007) focused on the length of fish at the fourth annulus, i.e. when fish were four years of age. From ages 1 to 4 years, cod in this population have low vulnerability to fishing, older cod (5 to 11 years) being much more likely to be caught by fishing gear.

The otolith data allowed Swain et al. (2007) to estimate the selection differential, S (sub-section 3.2.2), for a proxy of individual growth—the length of a fish at the fourth annulus. Let $L4$ be the length at the fourth annulus on the otolith. For each age i in the parental group (i ranged from 5 to 11 years) producing the offspring in cohort or year class j, the selection differential (S) for fish of age i is the difference in $L4$ between cod of age i in year j and the $L4$ of age 4 cod $i - 4$ years earlier. For example, for parents that are eight years old in year j, S equals the $L4$ of age 8 fish in year j (i.e. their back-calculated mean length at age 4) *minus* the $L4$ of four-year-olds in year $j - 4$ (see Figure 10.13(a)). For parents of age 5 in year j, S equals the back-calculated $L4$ of age 5 fish in year j minus the $L4$ of four-year-olds in year $j - 1$.

In the absence of selection, the length at the fourth annulus ($L4$) of an eight-year-old born in cohort j should be equal to the $L4$ in a four-year-old born in cohort j. However, if the average length at the fourth annulus is shorter in an eight-year-old born in cohort j than in a four-year-old from cohort j, this suggests that selection between ages four and

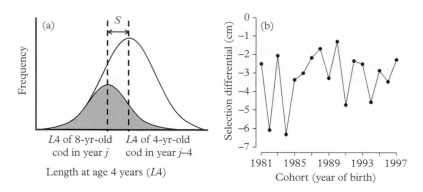

Figure 10.13 *Fishing has potential to select against faster growing individuals in a cod population in the southern Gulf of St. Lawrence, Canada. (a) A diagrammatic representation of how the selection differential (S) for growth is calculated for cod that are eight years old in year j. Here, S equals the L4 (length at age 4) of age 8 fish in year j minus the L4 of four-year-olds in year j − 4. (b) The selection differentials for 8-year-old cod born from 1981 to 1997.*

Data from Swain et al. (2007).

eight years of age has favoured slower growing fish, such that fish with shorter lengths at age four survived better to age eight than fish with longer lengths at age four.

The selection differentials were negative for all cohorts born from 1981 to 1997 (as an example, estimates for cod aged eight years are shown in Figure 10.13(b)). These negative values of S indicated greater survival to the parental, adult stage for slower-growing individuals. Swain et al. (2007) hypothesized that this differential survival was a consequence of fisheries-induced selection. The idea here is that faster growing individuals, being larger at a given age than slower growing individuals, are more likely to be caught by size-selective fisheries that target larger fishes, resulting in selection against fast growth and favouring slower growing individuals. Swain et al. (2007) estimated the heritability for length at age 4 to be (a somewhat high) 0.59. Given that a response to selection (R) can be expected when heritability (h^2) and selection differential (S) are both significantly different from zero ($R=h^2S$), their findings are consistent with the hypothesis that there had been genetic changes in growth in response to size-selective fishing mortality.

It is worth noting that these potential examples of FIE involved the selective removal of relatively large or small individuals. This has contributed to a misconception that exploitation (fishing or hunting) must be age- or size-selective for harvest-induced evolution to occur. It is important to recall from first principles (presented in Chapter 6) how changes in adult mortality, whether or not they are age- or size-selective, can be sufficient to generate evolutionary changes in life histories. That is, simple imposition of an unduly high level of mortality that is random with respect to size and age also has potential to cause harvest-induced evolution.

10.5.3   Hunting-induced evolution

Evidence of harvest-induced evolution is not restricted to fisheries. In western Canada, there are trophy hunts for bighorn sheep (*Ovis canadensis*). The longer the horn of the

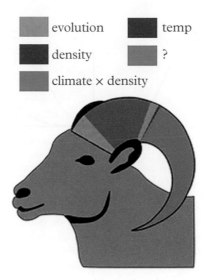

Figure 10.14 *Horn length of bighorn sheep is caused by different factors: genetic change (8.8%); density (26.5%); interaction between climate and density (2.9%); spring temperature (0.9%); and unexplained variation (60.9%).*

Re-drawn from Pelletier and Coltman (2018). Reprinted by permission from BMC.

male sheep, the greater the 'trophy', and the higher the probability of being shot. This size-selective hunt has contributed to reductions in horn length (a heritable trait, $h^2 = 0.36$; Pigeon et al. 2016) due, in part, to evolution (Coltman et al. 2003; Douhard et al. 2017) (Figure 10.14). Although horn length is not a life-history trait, it does influence reproductive success (Douhard et al. 2017) and, thus, b_x.

Evidence of a direct influence on life history is available from a study of heavily hunted brown bears (*Ursus arctos*) in Sweden. Based on 20 years of data, Van de Walle et al. (2018) concluded that a hunting regulation designed to protect females that have dependent young has potential to select for females that prolong their period of parental care. Between 1987 and 2004, 7.1 per cent (6 of 84 litters) of the young received 2.5 years of care as opposed to 1.5 years of care. However, between 2005 and 2015, the percentage of young receiving 2.5 years of care had increased five-fold (36.2 per cent; 29 of 80 litters). By extending their period of parental care, females trade off reduced future reproductive opportunities (cubs must be weaned before females can breed again) by increasing their survival and that of their young.

A third potential example of hunting-induced evolution in terrestrial animals comes from a 22-year study of wild boars (*Sus scrofa scrofa*) in the forest of Châteauvillain-Arc-en-Barrois in northeast France. Gamelon et al. (2011) found evidence of selection for earlier birth dates of boars subjected to high hunting pressure. Earlier birth date can come at a cost of reduced offspring survival; the earlier the birth date, the lower the likelihood that the birth coincides with favourable environmental conditions. But, under

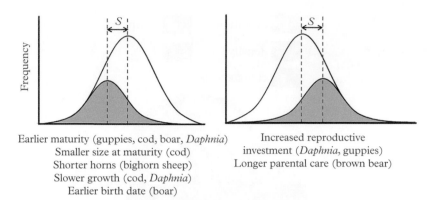

Earlier maturity (guppies, cod, boar, *Daphnia*)   Increased reproductive
Smaller size at maturity (cod)   investment (*Daphnia*, guppies)
Shorter horns (bighorn sheep)   Longer parental care (brown bear)
Slower growth (cod, *Daphnia*)
Earlier birth date (boar)

Figure 10.15 *An evolutionary response to selection can be estimated from the equation* $R = h^2S$. *A directional shift (decrease or increase) in the value of a trait requires that the trait be heritable* ($h^2 \neq 0$) *and that the selection differential (S) be significantly greater or less than zero. This figure illustrates differences in the frequency of trait values in the absence of selection (unshaded distributions) and after the imposition of selection (shaded distributions). The selection events are related to fishing and hunting (including predation). S is the difference in the mean value of the trait between the two frequency distributions. The panels reflect selection against larger trait values (left panel) and selection against smaller trait values (right panels). The examples are potential or demonstrated genetic responses to increased adult mortality discussed from Chapters 6 and 10.*

severe hunting pressure, which drastically reduces longevity, earlier birth can be favoured because it provides juvenile females with a potentially longer period of time for growth to occur. The longer the growth period, the greater the probability that females reach the threshold size of maturity by age 1, thus increasing the likelihood of having some reproductive success in an environment in which hunting-induced selection would select against delayed reproduction.

Humans are a dominant selective force, often causing more rapid phenotypic change in natural populations than other drivers (Alberti et al. 2017). The argument that sufficiently high exploitation mortality (selective or not) can cause genetic change in exploited populations has gained considerable traction since Handford et al.'s (1977) work on whitefish in the late 1970s. The intervening decades have provided compelling experimental and model-based evidence that harvest-induced evolution can cause genetic changes in life-history traits directly and indirectly through changes in behaviour (Biro and Post 2008; Van de Walle et al. 2018) and morphology (e.g. Douhard et al. 2017). This evidence includes temporal changes in probabilistic maturation reaction norms (Heino et al. 2015), laboratory selection experiments (Edley and Law 1988; Conover and Munch 2002; Uusi-Heikkilä et al. 2015), and mathematical modelling (Stokes et al. 1993; Ernande et al. 2004; Dunlop et al. 2009). All that is required is sufficiently high trait heritability, selection intensity, and time (Allendorf and Hard 2009) (Figure 10.15).

10.6   Summing Up

The concept of maximum sustainable yield (MSY) underpins national and international efforts to harvest animal populations in a manner that prevents over-exploitation and optimizes yield. A core determinant of MSY is r_{max}. This is the parameter through which the influence of life history is manifest in science advice on sustainability, r_{max} being ultimately determined by age-specific survival (l_x) and fecundity (b_x). The theory underlying MSY has its origins in basic models of density-dependent population growth. To a first approximation, these models predict that the population size at which growth or productivity is highest occurs when a population is at half of carrying capacity. However, when data allow for a more reliable estimate, this fraction of carrying capacity, φ, at which MSY is theoretically achievable is estimated to be greater than 0.5 for some species, such as whales and forest ungulates, but less than 0.5 for some marine fishes.

Another key element in achieving sustainable exploitation is understanding how natural (M) and harvest-induced (F) mortality changes with age. When data are of insufficient quality to allow φ, MSY, and M to be estimated by data-intensive approaches, simple (albeit increasingly uncertain) approximations can be used, some of which are based on the concept of life-history invariants introduced in Chapter 2. Complex and simple MSY-based methods can also be used in the setting of fishery management reference points related to population size and fishing mortality. Reference points define targets and limits to guide sustainable harvesting efforts.

When compared to the unexploited environments in which organismal life histories evolved, hunting and fishing impose additional sources of extrinsic mortality, often being several times greater than natural mortality, with direct consequences for l_x and potential influence on b_x. Thus, it is not surprising that hunting and fishing can affect life history. By selectively removing individuals on the basis of their size or age, or simply by imposing an unduly high level of mortality that is random with respect to size and age, exploitation has potential to generate evolutionary change. The potential importance of harvest-induced evolution to population viability and sustainable exploitation, relative to other factors known to affect the life history of exploited populations, remains a key question of scientific enquiry (Hutchings and Kuparinen 2020).

References

Abrams, P.A. (2000). The evolution of predator-prey interactions: theory and evidence. *Annual Review of Ecology and Systematics*, 31, 79–105.

Adler, P.H. and McCreadie, J.W. (2019). Black flies (Simuliidae). In G.R. Mullen and L.A. Durden, eds. *Medical and veterinary entomology*. 3rd edition, pp. 237–59. Academic Press, NY.

Albano, D.J. (1992). Nesting mortality of Carolina chickadees breeding in natural cavities. *Condor*, 94, 371–82.

Alberti, M., Correa, C., Marzluff, J.M., et al. (2017). Global urban signatures of phenotypic change in animal and plant populations. *Proceedings of the National Academy of Sciences of the U.S.A.*, 114, 8951–6.

Allendorf, F.W. and Hard, J.J. (2009). Human-induced evolution caused by unnatural selection through harvest of wild animals. *Proceedings of the National Academy of Sciences of the U.S.A.*, 106, 9987–94.

Alm, G. (1959). Connection between maturity, size, and age in fishes. Experiments carried out at the Kälarne Fishery Research Station. *Institute of Freshwater Research Drottningholm*, Report No. 40. Fishery Board of Sweden, Lund.

Alonzo, S.H. (2008). Conflict between the sexes and alternative reproductive tactics within a sex. In R.F. Oliveira, M. Taborsky and H.J. Brockmann, eds. *Alternative reproductive tactics*, pp. 435–50. Cambridge University Press, Cambridge, UK.

Alonzo, S.H. and Sinervo, B. (2001). Mate choice games, context-dependent good genes, and genetic cycles in the side-blotched lizard, *Uta stansburiana*. *Behavioral Ecology and Sociobiology*, 49, 176–86.

Andersen, K.H. (2019). *Fish ecology, evolution, and exploitation*. Princeton University Press, Princeton, NJ.

Andersson, S. and Shaw, R.G. (1994). Phenotypic plasticity in *Crepis tectorum* (Asteraceae): genetic correlations across light regimens. *Heredity*, 72, 113–25.

Archibald, S., Hempson, G.P. and Lehmann, C. (2019). A unified framework for plant life-history strategies shaped by fire and herbivory. *New Phytologist*, 224, 1490–503.

Arendt, J.D. (1997). Adaptive intrinsic growth rates: an integration across taxa. *Quarterly Review of Biology*, 72, 149–77.

Arnott, S.A., Chiba, S. and Conover, D.O. (2006). Evolution of intrinsic growth rate: metabolic costs drive trade-offs between growth and swimming performance in *Menidia menidia*. *Evolution*, 60, 1269–78.

Arnqvist, G. and Rowe, L. (2005). *Sexual conflict*. Princeton University Press, Princeton, NJ.

Arrow, G.H. (1951). *Horned beetles*. Junk, The Hague.

Australia (2018). *Guidelines for the Implementation of the Commonwealth Fisheries Harvest Strategy Policy*. 2nd edition. Department of Agriculture and Water Resources, Government of Australia, Canberra.

Baldwin, J.M. (1896). A new factor in evolution. *American Naturalist*, 30, 441–51.

Barneche, D.R., Robertson, D.R., White, C.R. and Marshall, D.J. (2018). Fish reproductive-energy output increases disproportionately with body size. *Science*, 360, 642–5.

Barnett, L.A.K., Branch, T.A., Ranasinghe, R.A. and Essington, T.E. (2017). Old-growth fishes become scarce under fishing. *Current Biology*, 27, 2843–8.

Barson, N.J., Aykanat, T., Hindar, K., et al. (2015). Sex-dependent dominance at a single locus maintains variation in age at maturity in salmon. *Nature*, 528, 405–8.

Barton, N.H., and Turelli, M. (1986). Evolutionary quantitative genetics: how little do we know? *Annual Review of Genetics*, 23, 337–70.

Beacham, T.D. and Murray, C.B. (1985). Effect of female size, egg size, and water temperature on developmental biology of chum salmon (*Oncorhynchus keta*) from the Nitinat River, British Columbia. *Canadian Journal of Fisheries and Aquatic Sciences*, 42, 1755–65.

Bell, G. (1976). On breeding more than once. *American Naturalist*, 110, 57–77.

Bell, G. (1980). The costs of reproduction and their consequences. *American Naturalist*, 116, 45–76.

Bennett, P.M. and Owens, I.P.F. (2002). *Evolutionary ecology of birds: life histories, mating systems, and extinction*. Oxford University Press, Oxford.

Benoît, H.P., Swain, D.P., Hutchings, J.A., Knox, D., Doniol-Valcroze, T. and Bourne, C.M. (2018). Evidence for reproductive senescence in a broadly distributed harvested marine fish. *Marine Ecology Progress Series*, 592, 207–24.

Beverton, R.J.H. (1963). Maturation, growth and mortality of Clupeid and Engraulid stocks in relation to fishing. *Rapports et proces-verbaux du Conseil internationale pour l'exploration de la Mer*, 154, 44–67.

Beverton, R.J.H. and Holt, S.J. (1959). A review of the lifespans and mortality rates of fish in nature and the relation to growth and other physiological characteristics. *CIBA Foundation Symposium—The Lifespan of Animals (Colloquia on Ageing)*, 5, 142–77.

Birch, L. (1948). The intrinsic rate of natural increase of an insect population. *Journal of Animal Ecology*, 1, 15–26.

Biro, P.A. and Post, J.R. (2008). Rapid depletion of genotypes with fast growth and bold personality traits from harvested fish populations. *Proceedings of the National Academy of Sciences of the U.S.A.*, 105, 2919–22.

Bleu, J., Gamelon, M. and Sæther, B.-E. (2016). Reproductive costs in terrestrial male vertebrates: insights from bird studies. *Proceedings of the Royal Society B*, 283: 20152600.

Boonekamp, J.J., Salomons, M., Bouwhuis, S., Dijkstra, C. and Verhulst, S. (2014). Reproductive effort accelerates actuarial senescence in wild birds: an experimental study. *Ecology Letters*, 17, 599–605.

Bose, A.P.H., Windorfer, J., Böhm, A., et al. (2020). Structural manipulations of a shelter resource reveal underlying preference functions in a shell-dwelling cichlid fish. *Proceedings of the Royal Society B*, 287. doi.org/10.1098/rspb.2020.0127.

Brakefield, P.M. and Zwaan, B.J. (2011). Seasonal polyphenisms and environmentally-induced plasticity in the Lepidoptera—the coordinated evolution of many traits on multiple levels. In T. Flatt and A. Heyland, eds. *Mechanisms of life history evolution: the genetics and physiology of life history traits and trade-offs*, pp. 243–52. Oxford University Press, Oxford, UK.

Brockelman, W.Y. (1975). Competition, fitness of offspring, and optimal clutch size. *American Naturalist*, 109, 677–99.

Brokordt, K.B., Guderley, H.E., Guay, M., Gaymer, C.F. and Himmelman, J.H. (2003). Sex differences in reproductive investment: maternal care reduces escape response capacity in the whelk *Buccinum undatum*. *Journal of Experimental Marine Biology and Ecology*, 291, 161–80.

Brommer, J.E. (2000). The evolution of fitness in life-history theory. *Biological Reviews*, 75, 377–404.

Brown, J.H., Gillooly, J.F., Allen, A.P., Savage, U.M. and West, G.B. (2004). Toward a metabolic theory of ecology. *Ecology*, 85, 1771–89.

Burger, J.R., Hou, C. and Brown, J.H. (2019). Toward a metabolic theory of life history. *Proceedings of the National Academy of the U.S.A.*, 116, 26,653–61.

Buzatto, B.A., Simmons, L.W. and Tomkins, J.L. (2012). Genetic variation underlying the expression of a polyphenism. *Journal of Evolutionary Biology*, 25, 748–58.

Cade, W.H. (1981). Alternative male strategies: genetic differences in crickets. *Science*, 212, 563–4.

Cade, W.H. and Cade, E.S. (1992). Male mating success, calling and searching behaviour at high and low densities in the field cricket, *Gryllus integer*. *Animal Behaviour*, 43, 49–56.

Calsbeek, R. and Sinervo, B. (2002). Uncoupling direct and indirect components of female choice in the wild. *Proceedings of the National Academy of Sciences of the U.S.A.*, 99, 14,897–902.

Capdevila, P., Stott, I., Beger, M. and Salguero-Gómez, R. (2020). Towards a comparative framework of demographic resilience. *Trends in Ecology and Evolution*, 35, 776–86.

Capinera, J.L. (1979). Qualitative variation in plants and insects: effect of propagule size on ecological plasticity. *American Naturalist*, 114, 350–61.

Castañeda, R.A., Burliuk, C.M.M., Casselman, J.M., et al. (2020). A brief history of fisheries in Canada. *Fisheries*, 45, 303–18.

Charlesworth, B. (1980). *Evolution in age-structured populations*. Cambridge University Press, Cambridge, UK.

Charlesworth, B. and León, J.A. (1976). The relation of reproductive effort to age. *American Naturalist*, 110, 449–59.

Charnov, E.L. (1982). *The theory of sex allocation*. Princeton University Press, Princeton.

Charnov, E.L. (1993). *Life history invariants*. Oxford University Press, Oxford.

Charnov, E.L., Gislason, H., and Pope, J.G. (2013). Evolutionary assembly rules for fish life histories. *Fish and Fisheries*, 14, 213–24.

Charnov, E.L. and Schaffer, W.M. (1973). Life-history consequences of natural selection: Cole's result revisited. *American Naturalist*, 107, 791–3.

Charnov, E.L., Warne, R. and Moses, M. (2007). Lifetime reproductive effort. *American Naturalist*, 170, E129–42.

Chin, A., Kyne, P.M., Walker, T.I., and McAuley, R.B. (2010). An integrated risk assessment for climate change: analysing the vulnerability of sharks and rays on Australia's Great Barrier Reef. *Global Change Biology*, 16, 1936–53.

Clausen, J., Keck, D.D. and Hiesey, W.M. (1948). *Experimental studies on the nature of species, volume III: environmental responses of climatic races of* Achillea. Publication 581. Carnegie Institution of Washington, Washington, DC.

Cody, M.L. (1966). A general theory of clutch size. *Evolution*, 20, 174–84.

Cohen, D. (1966) Optimizing reproduction in a randomly varying environment. *Journal of Theoretical Biology*, 12, 119–29.

Cole, L.C. (1954). The population consequences of life history phenomena. *Quarterly Review of Biology*, 29, 103–37.

Cole, S.C. (1990). Cod, God, country and family: the Portuguese-Newfoundland cod fishery. *Maritime Anthropological Studies*, 3, 1–29.

Collen, B., Dulvy, N.K., Gaston, K.J., et al. (2016). Clarifying misconceptions of extinction risk assessment with the IUCN Red List. *Biology Letters*, 12, 20150843. Available at: http://dx.doi.org/10.1098/rsbl.2015.0843.

Coltman, D.W., O'Donoghue, P., Jorgenson, J.T., Hogg, J.T., Strobeck, C. and Festa-Bianchet, M. (2003). Undesirable evolutionary consequences of trophy hunting. *Nature*, 426, 655–8.

Conover, D.O. (1992). Seasonality and the scheduling of life history at different latitudes. *Journal of Fish Biology*, 41, 161–78.

Conover, D.O. and Munch, S.B. (2002). Sustaining fisheries yields over evolutionary time scales. *Science*, 297, 94–6.

COSEWIC (2002). *COSEWIC assessment and update status report on the blue whale* Balaenoptera musculus *in Canada*. Committee on the Status of Endangered Wildlife in Canada, Ottawa. Available at: https://wildlife-species.canada.ca/species-risk-registry/virtual_sara/files/cosewic/sr_blue_whale_e.pdf.

COSEWIC (2007). *COSEWIC assessment and update status report on the rougheye rockfish* Sebastes sp. *in Canada*. Committee on the Status of Endangered Wildlife in Canada, Ottawa. Available at: https://wildlife-species.canada.ca/species-risk-registry/virtual_sara/files/cosewic/sr_sebastes_sp_e.pdf.

COSEWIC (2010). *COSEWIC assessment and update status report on the whitebark pine* Pinus albicaulis *in Canada*. Committee on the Status of Endangered Wildlife in Canada, Ottawa. Available at: https://wildlife-species.canada.ca/species-risk-registry/virtual_sara/files/cosewic/sr_Whitebark%20Pine_0810_e.pdf.

COSEWIC (2018). *COSEWIC assessment and update status report on the polar bear* Ursus maritimus *in Canada*. Committee on the Status of Endangered Wildlife in Canada, Ottawa. Available at: https://wildlife-species.canada.ca/species-risk-registry/virtual_sara/files/cosewic/OursBlancPolarBear-2019-Eng.pdf.

Courchamp, F., Berec, L. and Gascoigne, J. (2008). *Allee effects in ecology and conservation*. Oxford University Press, Oxford, UK.

Crow, J.F. (1958). Some possibilities for measuring selection intensities in man. *Human Biology*, 30, 1–13.

Darwin, C. (1859). *On the origin of species by means of natural selection*. John Murray, London.

Day, T. and Rowe, L. (2002). Developmental thresholds and the evolution of reaction norms for age and size at life-history transitions. *American Naturalist*, 159, 338–50.

de Magalhães, J.P. and Costa, J. (2009). A database of vertebrate longevity records and their relation to other life-history traits. *Journal of Evolutionary Biology*, 22, 1770–4.

Deevey, E.S. (1947). Life tables for natural populations of animals. *Quarterly Review of Biology*, 22, 283–314.

DeMaster, D., Angliss, R., Cochrane, J., et al. (2004). Recommendations to NOAA Fisheries ESA listing criteria by the quantitative working group. *NOAA Technical Memorandum* NMFS-F/SPO-67.

Dempster, E.R. and Lerner, I.M. (1950). Heritability of threshold characters. *Genetics*, 35, 212–36.

DFO. (2007). Assessing marine fish species: relating approaches based on reference points with approaches based on risk-of-extinction criteria. *Canadian Science Advisory Secretariat Proceedings Series 2007/024*, Department of Fisheries and Oceans, Ottawa.

Dillingham, P.W., Moore, J.E., Fletcher, D., et al. (2016). Improved estimation of intrinsic growth r_{max} for long-lived species: integrating matrix models and allometry. *Ecological Applications*, 26, 322–33.

Dobson, F.S. and Oli, M.K. (2008). The life histories of orders of mammals: fast and slow breeding. *Current Science*, 95, 862–5.

Dobzhansky, T. (1937). *Genetics and the origin of species*. Columbia University Press, NY.

Dobzhansky, T. (1950). Evolution in the tropics. *American Scientist*, 38, 209–21.

Douhard, M., Pigeon, G., Festa-Bianchet, M., Coltman, D.W., Guillemette, S. and Pelletier, F. (2017). Environmental and evolutionary effects on horn growth of male bighorn sheep. *Oikos*, 126, 1031–41.

Dulvy, N.K., Sadovy, Y. and Reynolds, J.D. (2003). Extinction vulnerability in marine populations. *Fish and Fisheries*, 4, 25–64.

Dulvy, N.K., Ellis, J.R., Goodwin, N.B., Grant, A., Reynolds, J.D. and Jennings, S. (2004). Methods of assessing extinction risk in marine fishes. *Fish and Fisheries*, 5, 255–76.

Dunlop, E.S., Enberg, K., Jørgensen, C. and Heino, M. (2009). Toward Darwinian fisheries management. *Evolutionary Applications*, 2, 245–59.

Eberhard, W.G. (1982). Beetle horn dimorphism: making the best of a bad job. *American Naturalist*, 119, 420–6.

Edley, M.T. and Law, R. (1988). Evolution of life histories and yields in experimental populations of *Daphnia magna*. *Biological Journal of the Linnean Society*, 34, 309–26.

Eis, S., Garman, E.H. and Ebell, L.F. (1965). Relation between cone production and diameter increment of Douglas fir (*Pseudotsuga menziesii* (Mirb.) Franco), grand fir (*Abies grandis* (Dougl.) Lindl.), and western white pine (*Pinus monticola* Dougl.). *Canadian Journal of Botany*, 43, 1553–9.

Emlen, D.J. (1996). Artificial selection on horn length-body size allometry in the horned beetle *Onthophagus acuminatus* (Coleoptera: Scarabaeidae). *Evolution*, 50, 1219–30.

Endler, J.A. (1986). *Natural selection in the wild*. Princeton University Press, Princeton, NJ.

Ernande, B., Dieckmann, U. and Heino, M. (2004). Adaptive changes in harvested populations: plasticity and evolution of age and size at maturation. *Proceedings of the Royal Society B*, 271, 415–23.

Fairbairn, D.J. and Yadlowski D.E. (1997). Coevolution of traits determining migratory tendency: correlated response of a critical enzyme, juvenile hormone esterase, to selection on wing morphometry. *Journal of Evolutionary Biology*, 10, 495–513.

Falconer, D.S. and Mackay, T.F.C. (1996). *Introduction to quantitative genetics*. 4th edition. Longmans Green, Harlow, Essex, UK.

FAO (2000). An appraisal of the suitability of the CITES criteria for listing commercially-exploited aquatic species. *FAO Fisheries Circular*. No. 954. Rome, Food and Agriculture Organization of the United Nations.

FAO (2018). *The state of world fisheries and aquaculture 2018 - Meeting the sustainable development goals*. Rome. Licence: CC BY-NC-SA 3.0 IGO.

Fenchel, T. (1974). Intrinsic rate of natural increase: the relationship with body size. *Oecologia*, 14, 317–26.

Fernie, A.R. and Yan, J. (2019). *De novo* domestication: an alternative route toward new crops for the future. *Molecular Plant*, 12, 615–31.

Field, D.L., Pickup, M. and Barrett, S.C.H. (2013). Comparative analyses of sex-ratio variation in dioecious flowering plants. *Evolution*, 67, 661–72.

Fisher, R.A. (1918). The correlation between relatives of the supposition of Mendelian inheritance. *Transactions of the Royal Society of Edinburgh*, 52, 399–433.

Fisher, R.A. (1930). *The genetical theory of natural selection*. Oxford University Press, Oxford.

Fisheries and Oceans Canada (2009). A fishery decision-making framework incorporating the precautionary approach. Available at: http://www.dfo-mpo.gc.ca/reports-rapports/regs/sff-cpd/precaution-eng.htm.

Flatt, T. (2014). Plasticity of lifespan: a reaction norm perspective. *Proceedings of the Nutrition Society*, 73, 532–42.

Fleming, I.A. (1996). Reproductive strategies of Atlantic salmon: ecology and evolution. *Reviews in Fish Biology and Fisheries*, 6, 379–416.

Foden, W.B., Butchart, S.H.M., Stuart, S.N., et al. (2013). Identifying the world's most climate change vulnerable species: a systematic trait-based assessment of all birds, amphibians and corals. *PLOS ONE*, 8, e65427.

Foden, W.B., Young, B.E., Akçakaya, H.R., et al. (2019). Climate change vulnerability assessment of species. *WIREs Climate Change*, 10(1). doi: 10.1002/wcc.551.

Folkvold, A., Jørgensen, C., Korsbrekke, K., Nash, R.D.M., Nilsen, T., Skjæraasen, J.E. (2014). Trade-offs between growth and reproduction in wild Atlantic cod. *Canadian Journal of Fisheries and Aquatic Sciences*, 71, 1106–12.

Fowler, C.W. (1988). Population dynamics as related to rate of increase per generation. *Evolutionary Ecology*, 2, 197–204.

Fowler, K. and Partridge, L. (1989). A cost of mating in female fruitflies. *Nature*, 338, 760–1.

Fox, M.G. and Keast, A. (1991). Effects of overwinter mortality on reproductive life history characteristics of pumpkinseed (*Lepomis gibbosus*) populations. *Canadian Journal of Fisheries and Aquatic Sciences*, 48, 1791–9.

Froese, R., Demirel, N., Coro, G., Kleisner, K.M. and Winker, H. (2017). Estimating fisheries reference points from catch and resilience. *Fish and Fisheries*, 18, 506–26.

Fryxell, J., Falls, J.B., Falls, E.A., Brooks, R.J., Dix, L. and Strickland, M. (2001). Harvest dynamics of mustelid carnivores in Ontario, Canada. *Wildlife Biology*, 7, 151–9.

Fryxell, J.M., Packer, C., McCann, K., Solberg, E.J. and Sæther, B.-E. (2010). Resource management cycles and the sustainability of harvested wildlife populations. *Science*, 328, 903–6.

Gadgil, M. and Bossert, W.H. (1970). Life historical consequences of natural selection. *American Naturalist*, 104, 1–24.

Gaillard, J.-M., Pontier, D., Allainé, D., Lebreton, J.-D., Trouvilliez, J. and Clobert, J. (1989). An analysis of demographic tactics in birds and mammals. *Oikos*, 56, 59–76.

Gamelon, M., Besnard, A., Gaillard, J.-M., et al. (2011). High hunting pressure selects for earlier birth date: wild boar as a case study. *Evolution*, 65, 3100–12.

Garner, S.R. and Neff, B.D. (2020). Alternative reproductive tactics. In R. Cothran and M. Thiel, eds. *Natural history of crustaceans: reproductive biology*. Oxford University Press, Oxford.

Gavrus-Ion, A., Sjøvold, T., Hernández, M., et al. (2017). Measuring fitness heritability: life history traits versus morphological traits in humans. *American Journal of Physical Anthropology*, 164, 321–30.

Gillespie, J.H. (1974). Natural selection for within-generation variance in offspring number. *Genetics*, 76, 601–6.

Gislason, H., Daan, N., Rice, J.C. and Pope, J.G. (2010). Size, growth, temperature and the natural mortality of marine fish. *Fish and Fisheries*, 11, 149–58.

Glick, P., Stein, B.A. and Edelson, N.A. (2011). *Scanning the conservation horizon: a guide to climate change vulnerability assessment*. National Wildlife Federation, Washington, DC.

Gompertz, B. (1820). A sketch of an analysis and notation applicable to the estimation of the value of life contingencies. *Philosophical Transactions of the Royal Society of London*, 110, 214–94.

Gompertz, B. (1825). On the nature of the function expressive of the law of human mortality, and on a new mode of determining the value of life contingencies. *Philosophical Transactions of the Royal Society of London*, 115, 513–83.

Gompertz, B. (1861). Supplement to two papers published in the Philosophical Transactions (1820 and 1825) on the science connected with human mortality. *Philosophical Transactions of the Royal Society of London*, 152, 511–59.

Goodman, D. (1984). Risk spreading as an adaptive strategy in iteroparous life histories. *Theoretical Population Biology*, 25, 1–20.

Grant, P.R. (1986). *Ecology and evolution of Darwin's finches*. Princeton University Press, Princeton, NJ.

Grantham, M.E., Antonio, C.J., O'Neil, B.R., Zhan, Y.X. and Brisson, J.A. (2016). A case for a joint strategy of diversified bet hedging and plasticity in the pea aphid wing polyphenism. *Biology Letters*, 12, 20160654. doi.org/10.1098/rsbl.2016.0654.

Grime, J.P. (1977). Evidence for the existence of three primary strategies in plants and its relevance to ecological and evolutionary theory. *American Naturalist*, 111, 1169–94.

Grime, J.P. and Pierce, S. (2012). *The evolutionary strategies that shape ecosystems.* Wiley-Blackwell, Oxford.

Gulland, J.A. (1971). *The fish resources of the ocean.* Fishing News Books, West Byfleet, UK.

Haeckel, R. (1866). *Generelle Morphologie der Organismen. Allgemeine Grundzige der organischen Formen-Wissenschaft, mechanisch begründet durch die von Charles Darwin reformirte Descendenz-Theorie.* 2 vols. Reimer, Berlin.

Haldane, J.B.S. (1937). The effects of variation on fitness. *American Naturalist*, 71, 337–49.

Haldorson, L. and Love, M. (1991). Maturity and fecundity in the rockfishes, *Sebastes* spp., a review. *Marine Fisheries Review*, 53, 25–31.

Hamel, S., Gaillard, J.-M., Yoccoz, N.G., Loisin, A., Bonenfant, C. and Descamps, S. (2010). Fitness costs of reproduction depend on life speed: empirical evidence from mammalian populations. *Ecology Letters*, 13, 915–35.

Hamilton, W.D. (1966). The moulding of senescence by natural selection. *Journal of Theoretical Biology*, 12, 12–45.

Handford, P., Bell, G. and Reimchen, T. (1977). A gillnet fishery considered as an experiment in artificial selection. *Journal of the Fisheries Research Board of Canada*, 34, 954–61.

Hare, J.A., Morrison, W.E., Nelson, M.W., et al. (2016). A vulnerability assessment of fish and invertebrates to climate change on the northeast U.S. continental shelf. *PLoS ONE*, 11, e0146756. doi:10.1371/journal.pone.0146756.

Harper, J.L. (1967). A Darwinian approach to plant ecology. *Journal of Ecology*, 55, 247–70.

Harper, J.L., Lovell P.H. and Moore K.G. (1970). The shapes and sizes of seeds. *Annual Review of Ecology and Systematics*, 1, 327–56.

Harvey, P.H. and Clutton-Brock, T.H. (1985). Life history variation in primates. *Evolution*, 39, 559–81.

Heino, M., Dieckmann, U. and Godø, H.R. (2002). Measuring probabilistic reaction norms for age and size at maturation. *Evolution*, 56, 669–78.

Heino, M., Pauli, B.D. and Dieckmann, U. (2015). Fisheries-induced evolution. *Annual Review of Ecology, Evolution, and Systematics*, 46, 461–80.

Hellens, R.P., Moreau, C., Lin-Wang, K., et al. (2010). Identification of Mendel's white flower character. *PLoS ONE*, 5, e13230.

Hilborn, R. (2001). Calculation of biomass trend, exploitation rate, and surplus production from survey and catch data. *Canadian Journal of Fisheries and Aquatic Sciences*, 58, 579–84.

Hilborn, R. (2020). Measuring fisheries management performance. *ICES Journal of Marine Science*, 77(7–8), 2432–8. doi:10.1093/icesjms/fsaa119.

Hirshfield, M.F. and Tinkle, D.W. (1975). Natural selection and the evolution of reproductive effort. *Proceedings of the National Academy of Sciences in the U.S.A.*, 72, 2227–31.

Hobday, A.J., Smith, A.D.M., Stobutzki, I.C., et al. (2011). Ecological risk assessment for the effects of fishing. *Fisheries Research*, 108, 372–84.

Hoenig, J.M. 1983. Empirical use of longevity data to estimate mortality rates. *Fishery Bulletin*, 82, 898–903.

Holt, S.J. (1958). The evaluation of fisheries resources by the dynamic analysis of stocks, and notes on the time factors involved. *ICNAF Special Publication*, 1, 77–95.

Hordyk, A.R. and Carruthers, T.R. (2018). A quantitative evaluation of a qualitative risk assessment framework: examining the assumptions and predictions of the Productivity Susceptibility Analysis (PSA). *PLoS ONE*, 13, e0198298.

Houde, A.E. and Endler, J.A. (1990). Correlated patterns of female mating preferences and male color patterns in the guppy *Poecilia reticulata. Science*, 248, 1405–8.

Hudson, E. and Mace, G. (1996). *Report of the workshop on marine fish and the IUCN Red List of Threatened Animals*. IUCN, Gland, CH. Available at: https://portals.iucn.org/library/node/11302.

Hutchings, J.A. (1994). Age- and size-specific costs of reproduction within populations of brook trout, *Salvelinus fontinalis. Oikos*, 70, 12–20.

Hutchings, J.A. (2000). Collapse and recovery of marine fishes. *Nature*, 406, 882–5.

Hutchings, J.A. (2001). Conservation biology of marine fishes: perceptions and caveats regarding assignment of extinction risk. *Canadian Journal of Fisheries and Aquatic Sciences*, 58, 108–21.

Hutchings, J.A. (2005). Life history consequences of overexploitation to population recovery in Northwest Atlantic cod (*Gadus morhua*). *Canadian Journal of Fisheries and Aquatic Sciences*, 62, 824–32.

Hutchings, J.A. (2011). Old wine in new bottles: reaction norms in salmonid fishes. *Heredity*, 106, 421–37.

Hutchings, J.A. (2015). Thresholds for impaired species recovery. *Proceedings of the Royal Society B*, 282, 20150654. Available at: http://dx.doi.org/10.1098/rspb.2015.0654.

Hutchings, J.A., Ardren, W.R., Barlaup, B.T., et al. (2019). Life-history variability and conservation status of landlocked Atlantic salmon: an overview. *Canadian Journal of Fisheries and Aquatic Sciences*, 76, 1697–708.

Hutchings, J.A. and Baum, J.K. (2005). Measuring marine fish biodiversity: temporal changes in abundance, life history and demography. *Philosophical Transactions of the Royal Society B-Biological Sciences*, 360, 315–38.

Hutchings, J.A., Butchart, S.H.M., Collen, B., Schwartz, M.K. and Waples, R.S. (2012). Red flags: correlates of impaired species recovery. *Trends in Ecology and Evolution*, 27, 542–6.

Hutchings, J.A. and Kuparinen, A. (2017). Empirical links between natural mortality and recovery in marine fishes. *Proceedings of the Royal Society B*, 284, 20,170,693.

Hutchings, J.A. and Kuparinen, A. (2020). Implications of fisheries-induced evolution for population recovery: refocusing the science and refining its communication. *Fish and Fisheries*, 21, 453–64.

Hutchings, J.A. and Morris, D.W. (1985). The influence of phylogeny, size and behaviour on patterns of covariation in salmonid life histories. *Oikos*, 45, 118–24.

Hutchings, J.A. and Myers, R.A. (1987). Escalation of an asymmetric contest: mortality resulting from mate competition in Atlantic salmon, *Salmo salar. Canadian Journal of Zoology*, 65, 766–8.

Hutchings, J.A. and Myers, R.A. (1994). The evolution of alternative mating strategies in variable environments. *Evolutionary Ecology*, 8, 256–68.

Hutchings, J.A. and Myers, R.A. (1995). The biological collapse of Atlantic cod off Newfoundland and Labrador: an exploration of historical changes in exploitation, harvesting technology, and management. In R. Arnason and L.F. Felt, eds. *The north Atlantic fishery: strengths, weaknesses, and challenges*, pp. 78–98. Institute of Island Studies, Charlottetown.

Hutchings, J.A., Myers, R.A., García, V.B., Lucifora, L.O. and Kuparinen, A. (2012). Life-history correlates of extinction risk and recovery potential. *Ecological Applications*, 22, 1061–7.

Hutchings, J.A. and Rangeley, R.W. (2011). Correlates of recovery for Canadian Atlantic cod (*Gadus morhua*). *Canadian Journal of Zoology*, 89, 386–400.

ICES. (2020a). Cod (*Gadus morhua*) in divisions 7.e–k (western English Channel and southern Celtic Seas). *International Council for the Exploration of the Sea*, Copenhagen. Available at: http://ices.dk/sites/pub/Publication%20Reports/Advice/2020/2020/cod.27.7e-k.pdf.

ICES. (2020b). Hake (*Merluccius merluccius*) in subareas 4, 6, and 7, and in divisions 3.a, 8.a–b, and 8.d, Northern stock (Greater North Sea, Celtic Seas, and the northern Bay of Biscay). *International Council for the Exploration of the Sea*, Copenhagen. Available at: http://ices.dk/sites/pub/Publication%20Reports/Advice/2020/2020/hke.27.3a46-8abd.pdf.

Janzen, D.H. (1969). Seed-eaters versus seed size, number, toxicity and dispersal. *Evolution*, 23, 1–27.

Janzen, D.H. (1976). Why bamboos wait so long to flower. *Annual Review of Ecology and Systematics*, 7, 347–91.

Jennings, S., Reynolds, J.D. and Mills, S.C. (1998). Life history correlates of responses to fisheries exploitation. *Proceedings of the Royal Society B*, 265, 333–9.

Jensen, A.L. (1996). Beverton and Holt life history invariants result from optimal trade-off of reproduction and survival. *Canadian Journal of Fisheries and Aquatic Sciences*, 53, 820–2.

Jensen, A.L. (2002). Maximum harvest of a fish population that has the smallest impact on population biomass. *Fisheries Research*, 57, 89–91.

Johannsen, W. (1911). The genotype conception of heredity. *International Journal of Epidemiology*, 43, 989–1000.

Jones, J.W. (1959). *The salmon*. Collins, London.

Jones, M.A. and Hutchings, J.A. (2001). The influence of male parr body size and mate competition on fertilization success and effective population size in Atlantic salmon. *Heredity*, 86, 675–84.

Jones, M.A. and Hutchings, J.A. (2002). Individual variation in Atlantic salmon fertilization success: implications for effective population size. *Ecological Applications*, 12, 184–93.

Jonsson, B. and Jonsson, N. (2011). *Ecology of Atlantic salmon and brown trout*. Springer, Dordrecht.

Jukema, J. and Piersma, T. (2006) Permanent female mimics in a lekking shorebird. *Biology Letters*, 2, 161–4.

Kamler, E. (1992). *Early life history of fish: an energetics approach*. Chapman and Hall, London.

Karlsson, P.S. (1994). The significance of internal nutrient cycling in branches for growth and reproduction of *Rhododendrum lapponicum*. *Oikos* 70, 191–200.

Keith, D.A., Akçakaya, H.R., Thuiller, W., et al. (2008). Predicting extinction risks under climate change: coupling stochastic population models with dynamic bioclimatic habitat models. *Biology Letters*, 4, 560–3.

Kindsvater, H.K., Mangel, M., Reynolds, J.D. and Dulvy, N.K. (2016). Ten principles from evolutionary ecology essential for effective marine conservation. *Ecology and Evolution*, 6, 2125–38.

Klomp, H. (1970). The determination of clutch-size in birds a review. *Ardea*, 58, 1–124.

Krebs, J.R. and Davies, N.B. (1981). *An introduction to behavioural ecology*. Sinauer, Sunderland, MA.

Kuparinen, A. and Hutchings, J.A. (2017). Genetic architecture of age at maturity can generate divergent and disruptive harvest-induced evolution. *Philosophical Transactions of the Royal Society B*, 372, 20,160,035.

Küpper, C., Stocks, M., Risse, J.E., et al. (2016). A supergene determines highly divergent male reproductive morphs in the ruff. *Nature Genetics*, 48, 79–83.

Lack, D. (1947a). The significance of clutch-size. *Ibis*, 89, 302–52.

Lack, D. (1947b). The significance of clutch-size in the partridge (*Perdix perdix*). *Journal of Animal Ecology*, 16, 19–25.

Lack, D. (1948). The significance of litter-size. *Journal of Animal Ecology*, 17, 45–50.

Lake, P.S. (2013). Resistance, resilience and restoration. *Ecological Management and Restoration*, 14, 20–4.

Lamichhaney, S., Fan, G., Widemo, F., et al. (2016). Structural genomic changes underlie alternative reproductive strategies in the ruff (*Philomachus pugnax*). *Nature Genetics*, 48, 84–8.

Lande, R. (1979). Quantitative genetic analysis of multivariate evolution, applied to brain: body size allometry. *Evolution*, 33, 402–16.

Lande, R. (1982). A quantitative genetic theory of life history evolution. *Ecology*, 63, 607–15.

Lande, R. (1993). Risks of population extinction from demographic and environmental stochasticity and random catastrophes. *American Naturalist*, 142, 911–27.

Lank, D.B., Smith, C.M., Hanotte, O., Burke, T. and Cooke, F. (1995) Genetic polymorphism for alternative mating behaviour in lekking male ruff *Philomachus pugnax*. *Nature*, 378, 59–62.

Law, R. (1979). The cost of reproduction in annual meadow grass. *American Naturalist*, 113, 3–16.

Law, R. (2007). Fisheries-induced evolution: present status and future directions. *Marine Ecology Progress Series*, 335, 271–7.

Le Quesne, W.J.F. and Jennings, S. (2012). Predicting species vulnerability with minimal data to support rapid risk assessment of fishing impacts on biodiversity. *Journal of Applied Ecology*, 49, 20–8.

Lee, R.D. (2003). Rethinking the evolutionary theory of aging: transfers, not births, shape senescence in social species. *Proceedings of the National Academy of the U.S.A.*, 100, 9637–42.

Leggett, W.C. and Carscadden, J.E. (1978). Latitudinal variation in reproductive characteristics of American shad (*Alosa sapidissima*): evidence for population specific life history strategies in fish. *Journal of the Fisheries Research Board of Canada*, 35, 1469–78.

Leishman, M.R., Wright, I.J., Moles, A.T., and Westoby, M. (2000). The evolutionary ecology of seed size. In M. Fenner, ed. *Seeds: the ecology of regeneration in plant communities*. CABI Publishing, Wallingford, UK.

Leivesley, J.A., Bussière, L.F., Pemberton, J.M., Pilkington, J.G., Wilson, K. and Hayward, A.D. (2019). Survival costs of reproduction are mediated by parasite infection in wild Soay sheep. *Ecology Letters*, 22, 1203–13.

Lepais, O., Manicki, A., Glise, S., Buoro, M. and Bardonnet, A. (2017). Genetic architecture of threshold reaction norms for male alternative reproductive tactics in Atlantic salmon (*Salmo salar* L.) *Scientific Reports*, 7, 43552. doi:10.1038/srep43552.

Lewontin, R.C. (1965). Selection for colonizing ability. In H.G. Baker and G.L. Stebbins, eds. *The genetics of colonizing species*, pp. 77–94. Academic Press, New York.

Loery, G., Pollock, K.H., Nichols, J.D. and Hines, J.E. (1987). Age-specificity of black-capped chickadee survival rates: analysis of capture-recapture data. *Ecology*, 68, 1038–44.

Lord, J.M. and Westoby, M. (2006). Accessory costs of seed production. *Community Ecology*, 150, 310–7.

Lotka, A.J. (1907). Relation between birth rates and death rates. *Science*, 26, 21–2.

Lowry, D.B. and Willis, J.H. (2010). A widespread chromosomal inversion polymorphism contributes to a major life history transition, local adaptation, and reproductive isolation. *PLoS Biology*, 8, e1000500.

Lush, J.L. (1937). *Animal breeding plans*. Collegiate Press, Ames, IA.

Lustenhouwer, N., Maynard, D.S., Bradford, M.A. et al. (2020). A trait-based understanding of wood decomposition by fungi. *Proceedings of the National Academy of Sciences of the U.S.A.*, 117, 11551–8.

MacArthur, R.H. and Wilson, E.O. (1967). *The theory of island biogeography*. Princeton University Press, Princeton, NJ.

Mace, G.M. and Lande, R. (1991). Assessing extinction threats: towards a reevaluation of IUCN threatened species categories. *Conservation Biology*, 5, 148–57.

Macfadyen, A. (1948). The meaning of productivity in biological systems. *Journal of Animal Ecology*, 17, 75–80.

Mahaut, L., Cheptou, P.-O., Fried, G., et al. (2020). Weeds: against the rules? *Trends in Plant Science*, 25, 1107–16.

Mahoney, N., Nol, E. and Hutchinson, T. (1997). Food-chain chemistry, reproductive success, and foraging behaviour of songbirds in acidified maple forests of southern Ontario. *Canadian Journal of Zoology*, 75, 509–17.

Malthus, T.R. (1798). *An essay on the principle of population*. J. Johnson, London.

Marshall, K.E. and Sinclair, B.J. (2010). Repeated stress exposure results in a survival-reproduction trade-off in *Drosophila melanogaster*. *Proceedings of the Royal Society B*, 277, 963–9.

Martin, T.G., Nally, S., Burbige, A.A., et al. (2012). Acting fast helps avoid extinction. *Conservation Letters*, 5, 274–80.

Matthews, S.N., Iverson, L.R., Prasad, A.M., Peters, M.P. and Rodewald, P.G. (2011). Modifying climate change habitat models using tree species-specific assessments of model uncertainty and life history-factors. *Forest Ecology and Management*, 262, 1460–72.

Maynard Smith, J. (1982). *Evolution and the theory of games*. Cambridge University Press, Cambridge, UK.

McCoy, M.W. and Gillooly, J.F. (2008). Predicting natural mortality rates of plants and animals. *Ecology Letters*, 11, 710–6.

McKitrick, M.C. (1993). Phylogenetic constraint on evolutionary theory: has it any explanatory power? *Annual Review in Ecology and Systematics*, 24, 307–30.

Medawar, P.B. (1952). *An unsolved problem of biology*. H. K. Lewis, London.

Michaletz, S.T., Weiser, M.D., Zhou, J., Kaspari, M., Helliker, B.R. and Enquist, B.J. (2015). Plant thermoregulation: energetics, trait-environment interactions, and carbon economics. *Trends in Ecology and Evolution*, 30, 714–24.

Michaud, E.J. and Echternacht, A.C. (1995). Geographic variation in the life history of the lizard *Anolis carolinensis* and support for the pelvic constraint model. *Journal of Herpetology*, 29, 86–97.

Michod, R.E. (1979). Evolution of life histories in response to age-specific mortality factors. *American Naturalist*, 113, 531–50.

Milner-Gulland, E.J. and Akçakaya, H.R. (2001). Sustainable indices for exploited populations. *Trends in Ecology and Evolution*, 16, 686–92.

Minto, C., Myers, R.A. and Blanchard, W. (2008). Survival variability and population density in fish populations. *Nature*, 452, 344–8.

Moles, A.T., Ackerly, D.D., Webb, C.O., Tweddle, J.C., Dickie, J.B. and Westoby, M. (2005). A brief history of seed size. *Science*, 307, 576–80.

Moles, A.T., Warton, D.I. and Westoby, M. (2003). Do small-seeded species have higher survival through seed predation than large-seeded species? *Ecology*, 84, 3148–61.

Moles, A.T. and Westoby, M. (2006). Seed size and plant strategy across the whole life cycle. *Oikos*, 113, 91–105.

Monaghan, P. and Nager R.G. (1997). Why don't birds lay more eggs? *Trends in Ecology and Evolution*, 12, 270–4.

Moreau, R.E. (1944). Clutch size: a comparative study, with special reference to African birds. *Ibis*, 86, 286–347.

Mousseau, T.A. and Roff, D.A. (1988). Natural selection and the heritability of fitness components. *Heredity*, 59, 181–97.

Moyle, P.B., Kiernan, J.D., Crain, P.K., Quiñones, R.M. (2013). Climate change vulnerability of native and alien freshwater fishes of California: a systematic assessment approach. *PLoS ONE*, 8, e63883. doi:10.1371/journal.pone.0063883.

Murphy, G. (1968). Pattern in life history and the environment. *American Naturalist*, 102, 391–403.

Musick, J.A. (1999). Criteria to define extinction risk in marine fishes. *Fisheries*, 24, 6–14.

Myers, R.A., Hutchings, J.A. and Barrowman, N.J. (1996). Hypotheses for the decline of cod in the North Atlantic. *Marine Ecology Progress Series*, 138, 293–308.

Myers, R.A., Hutchings, J.A. and Gibson, R.J. (1986). Variation in male parr maturation within and among populations of Atlantic salmon, *Salmo salar. Canadian Journal of Fisheries and Aquatic Sciences*, 43, 1242–8.

Myers, R.A., Mertz, G. and Fowlow, P.S. (1997). Maximum population growth rates and recovery times for Atlantic cod, *Gadus morhua. Fishery Bulletin*, 95, 762–72.

Myhrvold, N., Baldridge, E., Chan, B., Sivam, D., Freeman, D.L. and Ernest, S.K.M. (2015). An amniote life-history database to perform comparative analyses with birds, mammals, and reptiles. *Ecology*, 96, 3109.

Nee, S., Colegrave, N., West, S.A. and Grafen, A. (2005). The illusion of invariant quantities in life histories. *Science*, 309, 1236–9.

Neubauer, P., Jensen, O.P., Hutchings, J.A. and Baum, J.K. (2013). Resilience and recovery of overexploited marine populations. *Science*, 340, 347–9.

Nevoux, M., Forcada, J., Barbaud, C., Croxall, J. and Weimerskirch, H. (2010). Bet-hedging response to environmental variability, an intraspecific comparison. *Ecology*, 91, 2416–27.

Niel, C. and Lebreton, J.-D. (2005). Using demographic invariants to detect overharvested bird populations from incomplete data. *Conservation Biology*, 19, 826–35.

Nussey, D.H., Clutton-Brock, T.H., Albon, S.D., Pemberton, J. and Kruuk, L.E.B. (2005). Constraints on plastic responses to climate variation in red deer. *Biology Letters*, 1, 457–60.

Obeso, J.R. (2002). The costs of reproduction in plants. *New Phytologist*, 155, 321–48.

Olden, J.D., Hogan, Z.S. and Vander Zanden, M.J. (2007). Small fish, big fish, red fish, blue fish: size-based extinction risk of the world's freshwater and marine fishes. *Global Ecology and Biogeography*, 16, 694–701.

Oliveira, B., São-Pedro, V., Santos-Barrera, G., et al. (2017). AmphiBIO, a global database for amphibian ecological traits. *Scientific Data*, 4, 170,123. doi.org/10.1038/sdata.2017.123.

Oliveira, R.F., Taborsky, M. and Brockmann, H.J. (2008). *Alternative reproductive tactics.* Cambridge University Press, Cambridge, UK.

Ollivier, L. (2008). Jay Lush: reflections on the past. *Lohman Information*, 43, 3–12.

Oomen, R.A. and Hutchings, J.A. (2015). Variation in spawning time promotes genetic variability in population responses to environmental change in a marine fish. *Conservation Physiology*, 3, cov027.

Oomen, R.A. and Hutchings, J.A. (2020). Evolution of reaction norms. In D.J. Futuyma, ed. *Oxford bibliographies in evolutionary biology.* Oxford University Press, Oxford, UK. doi:10.1093/OBO/9780199941728-0130.

Oomen, R.A., Kuparinen, A., and Hutchings, J.A. (2020). Consequences of single-locus and tightly linked genomic architectures for evolutionary responses to environmental change. *Journal of Heredity*, 111, 319–32.

Pardo, S.A., Cooper, A.B. and Dulvy, N.K. (2013). Avoiding fishy growth curves. *Methods in Ecology and Evolution*, 4, 353–60.

Parker, G.A. (1984). Evolutionarily stable strategies. In J.R. Krebs and N.B. Davies, eds. *Behavioural ecology: an evolutionary approach*, pp. 30–61. Blackwell, Oxford.

Parker, G.A. and Begon, M. (1986). Optimal egg size and clutch size: effects of environment and maternal phenotype. *American Naturalist*, 128, 573–92.

Partridge, L. (1988). The rare-male effect: what is its evolutionary significance? *Philosophical Transactions of the Royal Society of London B*, 319, 525–39.

Partridge, L. and Harvey, P. (1988). The ecological context of life history evolution. *Science*, 241, 1449–55.

Partridge, L. and Sibly, R. (1991). Constraints in the evolution of life histories. *Philosophical Transactions of the Royal Society B*, 332, 3–13.

Patrick, W.S., Spencer, P., Link, J., et al. (2010). Using productivity and susceptibility indices to assess the vulnerability of United States fish stocks to overfishing. *Fishery Bulletin*, 108, 305–22.

Pauly, D. (1980). On the interrelationships between natural mortality, growth parameters, and mean environmental temperature in 175 fish stocks. *Journal du Conseil International pour l'Exploration de la Mer*, 39, 175–92.

Pauly, D. (2007). Ransom Aldrich Myers: chronicler of declining fish populations. *Nature*, 447, 160.

Pearl, R. and Miner, J.R. (1935). Experimental studies on the duration of life. XIV. The comparative mortality of certain lower organisms. *Quarterly Review of Biology*, 10, 60–79.

Pearson, R.G., Stanton, J.C., Shoemaker, K.T., et al. (2014). Life history and spatial traits predict extinction risk due to climate change. *Nature Climate Change*, 4, 217–21.

Pelletier, F. and Coltman, D.W. (2018). Will human influences on evolutionary dynamics in the wild pervade the Anthropocene? *BMC Biology*, 16. Available at: https://doi.org/10.1186/s12915-017-0476-1.

Pelletier, F., Page, K.A., Ostiguy, T. and Festa-Bianchet, M. (2005). Fecal counts of lungworm larvae and reproductive effort in bighorn sheep, *Ovis canadensis*. *Oikos*, 110, 473–80.

Pianka, E.R. (1970). On *r* and *K* selection. *American Naturalist*, 104, 592–7.

Pianka, E.R. (1979). This week's citation classic. *Citation Classic*, 47, 191.

Pianka, E.R. and Parker, W.S. (1975). Age-specific reproductive tactics. *American Naturalist*, 109, 453–64.

Piché, J., Hutchings, J.A. and Blanchard, W. (2008). Genetic variation in threshold reaction norms for alternative reproductive tactics in male Atlantic salmon, *Salmo salar*. *Proceedings of the Royal Society B*, 275, 1571–5.

Pierce, S., Negreiros, D., Cerabolini, B.E.L., et al. (2017). A global method for calculating plant CSR ecological strategies applied across biomes world-wide. *Functional Ecology*, 31, 444–57.

Pigeon, G., Festa-Bianchet, M., Coltman, D.W. and Pelletier, F. (2016). Intense selective hunting leads to artificial evolution in horn size. *Evolutionary Applications*, 9, 521–30.

Powles, H., Bradford, M.J., Bradford, R.G., et al. (2000). Assessing and protecting endangered marine species. *ICES Journal of Marine Science*, 57, 669–76.

Prince, J., Hordyk, A., Valencia, S.R., Loneragan, N. and Sainsbury, K. (2015). Revisiting the concept of Beverton-Holt life-history invariants with the aim of informing data-poor fisheries assessment. *ICES Journal of Marine Science*, 72, 194–203.

Prince, J.D., Wilcox, C. and Hall, N. (2021). Life history ratios: invariant or dimensionless ratios adapted to stoichiometric niches? *Fish and Fisheries*.

Promislow, D.E.L. and Harvey, P.H. (1990). Living fast and dying young: a comparative analysis of life-history variation among mammals. *Journal of Zoology*, 220, 417–37.

Punt, A.E., Smith, A.D.M., Smith, D.C., Tuck, G.N. and Klaer, N.L. (2014). Selecting relative abundance proxies for B_{MSY} and B_{MEY}. *ICES Journal of Marine Science*, 71, 469–83.

Quinn, T.P. (2018). *The behavior and ecology of Pacific salmon and trout*. 2nd edition. University of Washington Press, Seattle, WA.

Ramsay, S.M., Mennill, D.J., Otter, K.A., Ratcliffe, L.M. and Boag, P.T. (2003). Sex allocation in black-capped chickadees *Poecile atricapilla*. *Journal of Avian Biology*, 34, 134–9.

Realé, D., Garant, D., Humphries, M.M., Bergeron, P., Careau, V. and Montiglio, P.-O. (2010). Personality and the emergence of the pace-of-life syndrome concept at the population level. *Proceedings of the Royal Society B*, 365, 4051–63.

Redford, K.H., Amato, G., Baillie, J., et al. (2011). What does it mean to successfully conserve a (vertebrate) species? *BioScience*, 61, 39–48.

Regan, T., Taylor, B., Thompson, G., et al. (2009). Developing a structure for quantitative listing criteria for the U.S. Endangered Species Act using performance testing: Phase 1 report. NOAA Technical Memorandum NMFS-SWFSC-437.

Rensch, B. (1938). Einwirkung des Klimas bei der Ausprägung von Vogelrassen, mit besonderer Berücksichtigung der Flügelform und der Eizahl. In *Proceedings of the 8th International Ornithological Congress, Oxford, 1934*, pp. 285–311. Oxford.

Reynolds, J.D., Dulvy, N.K., Goodwin, N.B. and Hutchings, J.A. (2005). Biology of extinction risk in marine fishes. *Proceedings of the Royal Society B*, 272, 2337–44.

Reznick, D.N. (1985). Costs of reproduction: an evaluation of the empirical evidence. *Oikos*, 44, 257–67.

Reznick, D.N., Bryga, H. and Endler, J.A. (1990). Experimentally induced life-history evolution in a natural population. *Nature*, 346, 357–9.

Reznick, D.N., Shaw, F.H., Rodd, F.H. and Shaw, R.G. (2007). Evaluation of the rate of evolution in natural populations of guppies (*Poecilia reticulata*). *Science*, 275, 1934–7.

Ricker, W.E. (1981). Changes in the average size and average age of Pacific salmon. *Canadian Journal of Fisheries and Aquatic Sciences*, 38, 1636–56.

Ricklefs, R.E. and Wikelski, M. (2002). The physiology life-history nexus. *Trends in Ecology and Evolution*, 17, 462–8.

Rieseberg, L.H. (2001). Chromosomal rearrangements and speciation. *Trends in Ecology and Evolution* 16, 351–8.

Riley, M. (1999). Correlates of smallest sizes for microorganisms. In National Research Council, ed. *Size limits of very small microorganisms*, pp. 21–5. National Academy Press, Washington, DC.

Robinson, J.G. and Redford, K.H. (1991). Sustainable harvest of neotropical forest animals. In J.G. Robinson and K.H. Redford, eds. *Neotropical wildlife use and conservation*, pp. 415–29. University of Chicago Press, Chicago.

Rodrígues-Muñoz, R., Boonekamp, J.J., Liu, X.P., et al. (2018). Testing the effect of early-life reproductive effort on age-related decline in a wild insect. *Evolution*, 73, 317–28.

Roff, D.A. (1984). The evolution of life history parameters in teleosts. *Canadian Journal of Fisheries and Aquatic Sciences*, 41, 989–1000.

Roff, D.A. (1992). *Evolution of life histories: theory and analysis*. Chapman and Hall, New York.

Roff, D.A. (1996). The evolution of threshold traits in animals. *Quarterly Review of Biology*, 71, 3–35.

Roff, D.A. (2002). *Life history evolution*. Sinauer Associates, Sunderland.

Roff, D.A. and Mousseau, T.A. (1987). Quantitative genetics and fitness: lessons from *Drosophila*. *Heredity*, 58, 103–18.

Rollinson, N. and Hutchings, J.A. (2013). Environmental quality predicts optimal egg size in the wild. *American Naturalist*, 182, 76–90.

Rombough, P.J. (1985). Initial egg weight, time to maximum alevin wet weight, and optimal ponding times for Chinook salmon (*Oncorhynchus tshawytscha*). *Canadian Journal of Fisheries and Aquatic Sciences*, 42, 287–91.

Rose, M.R. (1984). Laboratory evolution of postponed senescence in *Drosophila melanogaster*. *Evolution*, 38, 1004–10.

Rose, M.R. and Charlesworth, B. (1981). Genetics of life history in *Drosophila melanogaster*. II. Exploratory selection experiments. *Genetics*, 97, 187–96.

Rowe, L. (1994). The costs of mating and mate choice in water striders. *Animal Behaviour*, 48, 1049–56.

Royauté, R., Anderson Berdal, M., Garrison, C.R. and Dochtermann, N.A. (2018). Paceless life? A meta-analysis of the pace-of-life syndrome hypothesis. *Behavioral Ecology and Sociobiology*, 72, 64. doi.org/10.1007/s00265-018-2472-z.

Russell, E.S. (1931). Some theoretical considerations on the 'overfishing' problem. *ICES Journal of Marine Science*, 6, 3–20.

Sadovy, Y. (2001). The threat of fishing to highly fecund fishes. *Journal of Fish Biology*, 59 (Supplement A), 90–108.

Sæther, B.-E. and Engen, S. (2015). The concept of fitness in fluctuating environments. *Trends in Ecology and Evolution*, 30, 273–81.

Schaffer, W.M. (1974a). Optimal reproductive effort in fluctuating environments. *American Naturalist*, 108, 783–90.

Schaffer, W.M. (1974b). Selection for optimal life histories: the effects of age structure. *Ecology*, 55, 291–303.

Schaffer, W.M. and Rosenzweig, M.L. (1977). Selection for optimal life histories. II: Multiple equilibria and the evolution of alternative reproductive strategies. *Ecology*, 58, 60–72.

Schijns, R., Froese, R., Hutchings, J.A. and Pauly, D. (2021). Five centuries of cod catches in eastern Canada. *ICES Journal of Marine Science*.

Schindler, D.E., Hilborn, R., Chasco, B., et al. (2010). Population diversity and the portfolio effect in an exploited species. *Nature*, 465, 609–12.

Schlichting, C.D. and Pigliucci, M. (1998). *Phenotypic evolution: a reaction norm perspective*. Sinauer, Sunderland, MA.

Schmalhausen, I.I. (1949). *Factors of evolution: the theory of stabilizing selection*, Blakiston, Philadelphia, PA.

Schwander, T., Libbrecht, R. and Keller L. (2014). Supergenes and complex phenotypes. *Current Biology*, 24, R288–94.

Sedgwick, A. (1909). The influence of Darwin on the study of animal embryology. In A.C. Seward, ed. *Darwin and modern science: essays in commemoration of the centenary of the birth of Charles Darwin and of the fiftieth anniversary of the publication of The Origin of Species*, pp. 175–84. Cambridge University Press, Cambridge, UK.

Seger, J. and Brockmann, H.J. (1987). What is bet-hedging? *Oxford Surveys in Evolutionary Biology*, 4, 182–211.

Seigel, R.A. and Fitch, H.S. (1984). Ecological patterns of relative clutch mass in snakes *Oecologia*, 61, 293–301.

Seigel, R.A., Fitch, H.S. and Ford, N.B. (1986). Variation in relative clutch mass in snakes among and within species. *Herpetologica*, 42, 179–85.

Sharpe, D.M.T. and Hendry, A.P. (2009). Life history change in commercially exploited fish stocks: an analysis of trends across studies. *Evolutionary Applications*, 2, 260–75.

Sharpe, F.R. and Lotka, A.J. (1911). A problem in age-distribution. *Philosophical Magazine*, 21, 435–8.

Sheriff, M.J., Krebs, C.J. and Boonstra, R. (2009). The sensitive hare: sublethal effects of predator stress on reproduction in snowshoe hares. *Journal of Animal Ecology*, 78, 1249–58.

Shine, R. (1980). 'Costs' of reproduction in reptiles. *Oecologia*, 46, 92–100.

Shine, R. and Greer, A.E. (1991). Why are clutch sizes more variable in some species than in others? *Evolution*, 45, 1696–706.

Shuster, S.M. (2008). The expression of crustacean mating strategies. In R.F. Oliveira, M. Taborsky and H.J. Brockmann, eds. *Alternative reproductive tactics*, pp. 224–50. Cambridge University Press, Cambridge, UK.

Shuster, S.M. and Wade, M.J. (1991). Equal mating success among male reproductive strategies in a marine isopod. *Nature*, 350, 608–10.

Shuster, S.M. and Wade, M.J. (2003). *Mating systems and strategies*. Princeton University Press, Princeton, NJ.

Shuter, B.J. and Post, J.R. (1990). Climate, population viability, and the zoogeography of temperate fishes. *Transactions of the American Fisheries Society*, 119, 314–36.

Silverton, J.W., Franco, M. and Harper, J.L. (1997). *Plant life histories: ecology, phylogeny, and evolution*. Cambridge University Press, Cambridge.

Simons, A.M. (2007). Selection for increased allocation to offspring number under environmental unpredictability. *Journal of Evolutionary Biology*, 20, 813–17.

Simons, A.M. (2009). Fluctuating natural selection accounts for the evolution of diversification bet hedging. *Proceedings of the Royal Society B*, 276, 1987–92.

Simons, A.M. (2011). Modes of response to environmental change and the elusive empirical evidence for bet hedging. *Proceedings of the Royal Society B*, 278, 1601–9.

Simons, A.M. and Johnston, M.O. (2006). Environmental and genetic sources of diversification in the timing of seed germination: implications for the evolution of bet hedging. *Evolution*, 60, 2280–92.

Sinervo, B. and Lively, C.M. (1996). The rock-paper-scissors game and the evolution of alternative male strategies. *Nature*, 380, 240–3.

Sinervo, B., Svensson, E. and Comendant, T. (2000). Density cycles and an offspring quantity and quality game driven by natural selection. *Nature*, 406, 985–8.

Sinervo, B. and Zamudio, K. (2001). The evolution of alternative reproductive strategies: fitness differential, heritability, and genetic correlation between the sexes. *Journal of Heredity*, 92, 198–205.

Slatkin, M. (1974). Hedging one's evolutionary bets. *Nature*, 250, 704–5.

Smith, C. and Wootton, R.J. (1995). The costs of parental care in teleost fishes. *Reviews in Fish Biology and Fisheries*, 5, 7–22.

Smith, C.C. and Fretwell, S.D. (1974). The optimal balance between size and number of offspring. *American Naturalist*, 108, 499–506.

Smith, S.M. (1995). Age-specific survival in breeding black-capped chickadees (*Parus atricapillus*). *Auk*, 12, 840–6.

Sparholt, H., Bogstad, B., Christensen, V., et al. (2020). Estimating F_{msy} from an ensemble of data sources to account for density dependence in Northeast Atlantic fish stocks. *ICES Journal of Marine Science*, 78, 55–69.

Stearns, S.C. (1976). Life-history tactics: a review of the ideas. *Quarterly Review of Biology*, 51, 3–47.

Stearns, S.C. (1983). The influence of size and phylogeny on patterns of covariation among life-history traits in the mammals. *Oikos*, 41, 173–87.

Stearns, S.C. (1992). *The evolution of life histories*. Oxford University Press, Oxford.

Stearns, S.C. and Koella, J.C. (1986). The evolution of phenotypic plasticity in life-history traits: predictions of reaction norms for age and size at maturity. *Evolution*, 40, 893–913.

Stephenson, R. L., Wiber, M., Paul, S., et al. (2019). Integrating diverse objectives for sustainable fisheries in Canada. *Canadian Journal of Fisheries and Aquatic Sciences*, 76, 480–96.

Stokes, T.K. and Law, R. (2000). Fishing as an evolutionary force. *Marine Ecology Progress Series*, 208, 307–9.

Stokes, T.K., McGlade, J.M. and Law, R. (1993). *The exploitation of evolving resources*. Springer-Verlag, Berlin.

Stresemann, E. (1934). Sauropsida: Aves. In W. Kükenthal and T., Krumbach, eds. *Handbuch der zoologie: eine naturgeschichte der stämme des tierreiches*. Walter de Gruyter & Co., Berlin.

Sultan, S.E. and Bazzaz, F.A. (1993). Phenotypic plasticity in *Polygonum persicaria*. II. Norms of reaction to soil moisture, ecological breadth, and the maintenance of genetic diversity. *Evolution*, 47, 1032–49.

Svärdson, G. (1949). Natural selection and egg number in fish. *Report of the Institute of Freshwater Research, Drottningholm*, 29, 115–22.

Swain, D.P. (2011). Life-history evolution and elevated natural mortality in a population of Atlantic cod (*Gadus morhua*). *Evolutionary Applications*, 4, 18–29.

Swain, D.P., Sinclair, A.F. and Hanson, J.M. (2007). Evolutionary response to size-selective mortality in an exploited fish population. *Proceedings of the Royal Society B*, 274, 1015–22.

Taborsky, M. (2008). Alternative reproductive tactics in fish. In R.F. Oliveira, M. Taborsky and H.J. Brockmann, eds. *Alternative reproductive tactics*, pp. 251–99. Cambridge University Press, Cambridge, UK.

Taborsky, M., Oliveira, R.F. and Brockmann, H.J. (2008). The evolution of alternative reproductive tactics: concepts and questions. In R.F. Oliveira, M. Taborsky and H.J. Brockmann, eds. *Alternative reproductive tactics*, pp. 1–21. Cambridge University Press, Cambridge, UK.

Taggart, J.B., McLaren, I.S., Hay, D.W., Webb, J.H. and Youngson, A.F. (2001). Spawning success in Atlantic salmon (*Salmo salar* L.): a long-term DNA profiling-based study conducted in a natural stream. *Molecular Ecology*, 10, 1047–60.

Then, A.Y., Hoenig, J.M., Hall, N.G. and Hewitt, D.A. (2015). Evaluating the predictive performance of empirical estimators of natural mortality rate using information on over 200 fish species. *ICES Journal of Marine Science*, 72, 82–92.

Thorson, J.T., Munch, S.B., Cope, J.M. and Gao, J. (2017). Predicting parameters for all fishes worldwide. *Ecological Applications*, 27, 2262–76.

Tidière, M., Gaillard, J.-M., Müller, D.W.H., et al. (2015). Does sexual selection shape sex differences in longevity and senescence patterns across vertebrates? A review and new insights from captive ruminants. *Evolution*, 69, 3123–40.

Tinkle, D.W. (1969). The concept of reproductive effort and its relation to the evolution of life histories in lizards. *American Naturalist*, 103, 501–16.

Tomkins, J.L. and Brown, G.S. (2004). Population density drives the local evolution of a threshold dimorphism. *Nature*, 431, 1099–103.

Trippel, E.A. (1995). Age at maturity as a stress indicator in fisheries. *Bioscience*, 45, 759–71.

Tuttle, E.M., Bergland, A.O., Korody, M.L., et al. (2016). Divergence and functional degradation of a sex chromosome-like supergene. *Current Biology*, 26, 344–50.

Twyford, A.D and Friedman, J. (2015). Adaptive divergence in the monkey flower *Mimulus guttatus* is maintained by a chromosomal inversion. *Evolution*, 69, 1476–86.

Uusi-Heikkilä, S., Whiteley, A.R., Kuparinen, A., et al. (2015). The evolutionary legacy of size-selective harvest extends from genes to populations. *Evolutionary Applications*, 8, 597–620.

USFWS (2016). Available at: https://www.fws.gov/refuge/Charles_M_Russell/what_we_do/science/BFF_recovery.html

Van de Walle, J., Pigeon, G., Zedrosser, A., Swenson, J.E. and Pelletier, F. (2018). Hunting regulation favors slow life histories in a large carnivore. *Nature Communications*, 9, 1100 doi:10.1038/s41467-018-03506-3.

Vladić, T. and Petersson, E. (2015). *Evolutionary biology of Atlantic salmon*. CRC Press, Boca Raton, FL.

von Bertalanffy, L. (1938). A quantitative theory of organic growth (inquiries on growth laws. II). *Human Biology*, 10, 181–213.

Wade, M.J. and Arnold, S.J. (1980). The intensity of sexual selection in relation to male sexual behaviour, female choice, and sperm precedence. *Animal Behaviour*, 28, 446–61.

Walker, S.E. and Cade, W.H. (2003). A simulation model of the effects of frequency dependence, density dependence and parasitoid flies on the fitness of male field crickets. *Ecological Modelling*, 169, 119–30.

Walker, T.J. (1986). Stochastic polyphenism: coping with uncertainty. *The Florida Entomologist*, 69, 46–62.

Walters, C. and Martell, S.J.D. (2002). Stock assessment needs for sustainable fisheries management. *Bulletin of Marine Science*, 70, 629–38.

Wang, H.-Y., Botsford, L.W., White, J.W., et al. (2014). Effects of temperature on life history sensitivity to fishing in Atlantic cod *Gadus morhua*. *Marine Ecology Progress Series*, 514, 217–29.

Waples, R.S. and Audzijonyte, A. (2016). Fishery-induced evolution provides insights into adaptive responses of marine species to climate change. *Frontiers in Ecology and the Environment*, 14, 217–24.

Waples, R.S., Nammack, M., Cochrane, J.F. and Hutchings, J.A. (2013). A tale of two acts: endangered species listing practices in Canada and the United States. *BioScience*, 63, 723–34.

Watson, W. and Walker, H.J. (2004). The world's smallest vertebrate, *Schindleria brevipinguis*, a new paedomorphic species in the family Schindleriidae (Perciformes: Gobioidei). *Records of the Australian Museum*, 56, 139–42.

Wedell, N. and Karlsson B. (2003). Paternal investment directly affects female reproductive effort in an insect. *Proceedings of the Royal Society B*, 270, 2065–71.

Weinbaum, K.Z., Brashares, J.S., Golden, C.D. and Getz, W.M. (2013). Searching for sustainability: are assessments of wildlife harvests behind the times? *Ecology Letters*, 16, 99–111.

Weir, L.K., Grant, J.W.A. and Hutchings, J.A. (2010). Patterns of aggression and operational sex ratio within alternative male phenotypes in Atlantic salmon. *Ethology*, 116, 166–75.

Wenk, E.H. and Falster, D.S. (2015). Quantifying and understanding reproductive allocation schedules in plants. *Ecology and Evolution*, 5, 5521–38.

West, S.A., Reece, S.E. and Sheldon BC (2002). Sex ratios. *Heredity*, 88, 117–24.

White, G.C. and Burnham, K.P. (1999). Program MARK: survival estimation from populations of marked animals. *Bird Study*, 46:sup1, S120–39. doi:10.1080/00063659909477239.

Wickman, P.-O. and Karlsson, B. (1989). Abdomen size, body size and the reproductive effort of insects. *Oikos*, 56, 209–14.

Wikelski, M., Spinney, L., Schelsky, W., Scheuerlein, A. and Gwinner, E. (2003). Slow pace of life in tropical sedentary birds: a common-garden experiment on four stonechat populations from different latitudes. *Proceedings of the Royal Society B*, 270, 2383–8.

Williams, G.C. (1957). Pleiotropy, natural selection, and the evolution of senescence. *Evolution*, 11, 398–411.

Williams, G.C. (1966). Natural selection, the costs of reproductions, and a refinement of Lack's principle. *American Naturalist*, 100, 687–90.

Wilson, A.N. and Thompson K. (1989). A comparative study of reproduction isolation in 40 British grasses. *Functional Ecology*, 3, 297–302.

Winterbottom, R. (1990). The *Trimmatom nanus* species complex (Actinopterygii, Gobiidae): phylogeny and progenetic heterochrony. *Systematic Zoology*, 39, 253–65.

Woltereck, R. (1909). Weitere experimentelle Untersuchungen über Artveränderung, speziell über das Wesen quantitativer Artunterschiede bei Daphniden. *Verhandlungen der Deutschen Zoologischen Gesellschaft*, 1909, 110–72.

Wootton, R.J. (1998). *Ecology of teleost fishes*. Chapman & Hall, London.

Young, T.P. (1981). A general model of comparative fecundity for semelparous and iteroparous life histories. *American Naturalist*, 118, 27–36.

Zamudio, K.R. and Chan, L.M. (2008). Alternative reproductive tactics in amphibians. In R.F. Oliveira, M. Taborsky and H.J. Brockmann, eds. *Alternative reproductive tactics*, pp. 300–31. Cambridge University Press, Cambridge, UK.

Zamudio, K.R. and Sinervo, B. (2000). Polygyny, mate guarding, and posthumous fertilizations as alternative male mating strategies. *Proceedings of the National Academy of Sciences of the United States of America*, 97, 14,427–32.

Zhou, S., Griffiths, S.P. and Miller, M. (2009). Sustainability assessment for fishing effects (SAFE) on highly diverse and data-limited fish bycatch in a tropical prawn trawl fishery. *Marine and Freshwater Research*, 60, 563–70.

Subject Index

Taxonomic Index